AF255716

The Musicality Within

A Scientific Perspective on Music Obsession

Chris Young Kelly

Contents

Chapter 1

The Mysterious Music

The "Big" Puzzle

Music is one of the most captivating and mysterious elements of human existence. It has been woven into the fabric of our lives for thousands of years, perhaps as long as, or even longer than, language. From daily routines and rituals to art, education, religion, and entertainment, music permeates every aspect of human culture. This deeply rooted connection shows up across many areas, including our daily lives, lifestyles, history, art, education, religion, business, and entertainment. The widespread presence of music in human culture has sparked our curiosity, leading us to question its nature, purpose, and the unique role it plays in shaping our emotions, behaviors, and interactions.

The possible questions span a broad spectrum, from literary and artistic to philosophical, theological, and scientific. These questions can range from the simple —such as what music is—to the complex, like exploring why we are so obsessed with it. Within this range, there are questions related to our everyday concerns as well as those that are more scientifically oriented. The former captures the interest of the general public and often leads to common-sense answers through popular search engines, Wikipedia, and AI. However, for those more curious individuals who seek deeper understanding, greater satisfaction is usually found in the domains of science and neuroscience. Interestingly, even questions that are less scientific in nature often turn to science for more satisfying answers. Whether questions are based on common sense or scientific inquiry, satisfying our curiosity about music remains a comprehensive scientific pursuit, regardless of how the questions are posed.

Science-oriented questions often aim to explain musical phenomena and their impact on individuals who engage with music by listening or performing. However, among the many intriguing questions, none is more fundamental than this: "Why" and "How" are humans so captivated by music? This scientific curiosity is twofold: it seeks to uncover the reasons behind our deep connection with music, which influences every aspect of our lives. A reasonable answer is that engaging in musical activities brings us a sense of satisfaction and pleasure. The follow-up and second part address how this pleasurable emotion arises. More specifically, how can science elucidate the joy of engaging with music?

Over the past century, these questions have engaged much of our intellectual capacity as experts from various fields, including philosophy, science, musicology, psychology, history, anthropology, neuroscience, chemistry, and physics, who have sought to resolve them. We have undoubtedly made progress on specific aspects of music, reaching a basic level of understanding thanks to dedicated scientific efforts. Yet, despite advancements in individual disciplines, they often remain siloed and disjoint. This book aims to show that nearly all musical questions ultimately require scientific explanations. Additionally, we will examine the connections among these scientific answers through a unifying principle that links complex scientific responses across diverse disciplines such as physics, neuroscience, and psychology.

Let's start by examining some common questions and their answers, such as what music means, what makes someone a musician, how music relates to culture, the role of music in education, and whether music is even necessary. This list is not intended to be comprehensive; instead, it highlights the book's focus on the scientific perspective on music. Although these questions may seem different, they share a common theme that, when explored across fields, can help explain why we are so fascinated with music.

The Frequently Asked Questions

What Is Music?

What is music? This seemingly simple question has sparked deep reflection throughout history. It is widely accepted that music is a structured arrangement of sounds and tones, made vocally or with instruments, that triggers emotional or physical responses. It acts as a form of expression that crosses cultures, generations, and locations, often stirring feelings from joy to sadness and from calm to tension. This musical arrangement is not just a collection of sounds but a dynamic system where rhythm, melody, harmony, and timbre blend to create something much greater than the sum of its parts. It communicates and connects, providing a unique messaging system that goes beyond the limits of words. In other words, it offers an emotional experience that regular languages cannot, because music engages many neural regions, offering complex instructions for the brain's reward system.

Although definitions vary, one thing is certain: music is a human creation, shaped by both intelligence and emotion, and found in every culture known to us. However, we can usually identify the cultural origins of the music we hear fairly easily. For example, I can easily tell Beethoven's 5th Symphony apart from India's Hanuman Chalisa. I can also quickly tell the difference between American and Chinese pop music. Still, beneath these differences, there are common elements: scales, melodies, rhythms, and harmonies that tend to look quite similar across different cultures.

What can explain this paradox of diversity and similarity in reference to the definition of music? The key to the answer is rooted in our biology and the underlying physics, which will become clear as we work through the book.

The Evolving Music Tastes

Musical tastes change over time, but music's emotional power stays the same across different generations. Today, many young people proudly call themselves Swifties, devoted fans of Taylor Swift. A few decades ago, the baby boomers danced to the Beach Boys and other pop icons of the 1960s. Older generations still remember the excitement of big-band music. Classical music fans, though few, continue to be captivated by the works of Mozart, Beethoven, and Brahms.

Every generation tends to look back at its music with fondness and view newer styles with suspicion. In the 1940s, parents were wary of Frank Sinatra. In the 1950s, they were scandalized by rock 'n' roll. By the 1970s, it was Johnny Cash and Elvis Presley who raised eyebrows. This cycle continues today. Older listeners often claim that contemporary music has lost its quality. And yet, every generation produces artists and music that leave lasting impacts.

What's happening isn't just about cultural nostalgia. It shows how we grow up surrounded by certain musical styles. Familiarity creates affection. People tend to love the music they heard in their youth; those emotional bonds are strong.

Yet, despite having very different tastes, something amazing happens: people come together in sync. We sing along, tap our feet, or dance to the beat, regardless of the genre. Even audiences at classical concerts, usually known for their quiet formality, become noticeably lively during André Rieu's elegant and energetic performances.

So, what motivates us to move and synchronize with music, regardless of genre, style, and time? Is there a deeper, shared mechanism in the human brain that causes this response? This craving for synchrony across generations and cultures indicates that music taps into something inherently human. Chapter 8 explores this further from a neuroscientific viewpoint, aiming to answer this question.

<u>How Music And Culture Relate</u>

We are increasingly becoming global citizens in the 21st century, and most of us have had the opportunity to experience various cultures, languages, and music. On the surface, music varies greatly across cultures due to a mix of historical, social, geographical, and psychological factors that influence how people create, perform, and experience music. Each culture develops its musical traditions, shaped by its unique environment, language, religion, and values.

For example, Western music often emphasizes harmony and structured forms; think of Beethoven's or Brahms's symphonies. Ancient Chinese classical music focuses on intricate improvisation and the interplay of multiple instruments. African music often emphasizes rhythm, polyrhythmic drumming, and call-and-response vocal traditions, which are influenced by communal, ritualistic, and spiritual practices.

Geography and languages further influence musical styles. Eastern music often uses a different subset of scales that subtly differ from Western norms, creating tonal textures that sound unfamiliar to Western listeners. Tonal languages, like Mandarin, naturally shape musical phrasing and melody. Additionally, Arabic music, with its use of microtones, reflects the flowing, expressive qualities of the Arabic language itself.

When learned ethnomusicologists study how scales, the foundation of melodies, were created and enjoyed in different cultures, they discover that all scales are based on similar concepts and principles. How did our ancestral musicians create nearly the same scales regardless of their cultural origins? Neuroscientific studies suggest this is not accidental; neural functions common to all humans lead to these unavoidable similarities.

Chapter 3 explains this phenomenon from a historical view, while Chapters 6 and 7 explore the neurological aspects. In both cases, we'll see how the universal recognition of the octave isn't a coincidence but reflects our shared neural functions and, by extension, humanity.

Culture influences its music, creating a distinct specificity that sets it apart from others. Yet, deep down, there is an uncanny commonality that can be traced back to biology.

<u>How Music and Religion Relate</u>

Throughout history, music has been central to religious life. Its emotional power has long been used to connect worshippers with the divine, foster shared experiences, and deepen spiritual meaning. In nearly every culture, religious leaders—whether shamans, priests, or monks—have used music as a tool for communication, ritual, and transformation.

Music's impact on religions is both practical and profound. Choral singing, communal hymns, and devotional chants build a sense of unity, belonging, and shared purpose. Music helps congregations feel connected and move together. In many traditions, it also acts as a way to connect with ancestral spirits, as seen in drumming, dancing, or sung prayers in indigenous or African diasporic practices.

Across the world's faiths, music acts as a bridge linking the sacred, people, communities, and individual feelings with collective beliefs.

Its ability to evoke awe, comfort, and transcendence makes it a perfect way for spiritual expression.

But where does this power come from? Some believe music's spiritual strength points to a divine source. Others argue that its emotional impact is rooted in biology, not metaphysics. Neuroscience has provided causal proof for the latter. Brain studies show that music activates circuits involved in emotion, memory, and reward, whether it is used in a sacred or secular setting.

In this view, religions have simply tapped into music's natural ability to move us. The same neural systems that cause us to feel spiritually uplifted during a hymn are activated when we feel emotionally stirred by a favorite song or a stadium concert. In fact, the fandom and almost-religious devotion seen in modern pop culture reflect the exact psychological mechanisms.

Chapter 8 will explore how these shared emotional circuits work—and how music can evoke feelings of unity and transcendence, whether in a temple or at a Taylor Swift concert.

<u>What Makes A Musician</u>

What makes musicians different? What enables some people to craft music that touches others so profoundly? Are they born with unique talents, or are early experiences and consistent practice what shape them?

We often picture musicians as rare, inspired figures, maybe even touched by something divine. They seem to perceive or feel music in a unique way. Some appear to compose with ease, while others struggle over each phrase, yet both can create equally powerful results. Where does that creative spark originate?

Today, we have the scientific tools and skills to start answering these questions. For the first time in human history, we can examine musical ability not just as an art form but as a product of the brain, at the intersection of biology, exposure, and cognitive development.

Research increasingly shows that humans are born with a basic capacity for music; we define this capacity as musicality in Chapter 2. We are all, in some way, musical. Most people can follow a beat, recognize a melody, or feel moved by a particular song. So why do some people become composers or performers, while others remain passive listeners?

The answer seems to depend on exposure and neural plasticity. The brain's capacity to adapt and reorganize itself, especially during early childhood, means that those who are immersed in music from a young age tend to develop stronger musical instincts. It's not magic or fate. It's biology working with the environment. Being a musician is simply a matter of exposure.

In later chapters, especially Chapters 6, 7, and 8, we will explore the neuroscience behind this. But the main idea is simple: the difference between musicians and non-musicians may be less about talent and more about experience, timing, and the brain's ability to change. Chapter 2 will support this by showing how deeply music influences our lives, past and present, how it quietly shapes who we are, and even how the world functions today, whether we see ourselves as musical or not.

<u>Can Music Make Us Smarter?</u>

This question has undoubtedly sparked people's imagination. For many years, the so-called "Mozart Effect" has suggested that listening to Mozart can boost intelligence, especially in children. While the science behind this idea has often been misunderstood and exaggerated, there is some truth: music does influence the brain in meaningful ways.

Listening to Mozart can temporarily improve specific tasks, especially those involving spatial reasoning. However, this does not equate to long-term cognitive development. Genuine educational benefits come not from passive listening but from actively learning music, playing an instrument, reading notation, or singing in a group.

Research consistently shows that music education improves a wide range of cognitive skills. It enhances memory, boosts pattern recognition, and supports abilities in language, math, and reading. Learning music also encourages critical thinking, problem-solving, and creativity.

But the benefits aren't just intellectual. Music education also fosters emotional growth. It offers students tools for self-expression, builds confidence and discipline, and encourages teamwork through ensemble performances. Students who study music often develop patience, resilience, and better time management skills that benefit them beyond the music classroom. There's also a cultural element. Through music, students experience different histories and traditions. They learn to appreciate various perspectives, promoting empathy and global awareness.

All of these points highlight something more fundamental: music education changes the brain. It rewires how we perceive the world, connect with others, and understand ourselves. What mechanisms drive these changes? How does learning music reshape the brain's biological, intellectual, and emotional makeup? Later chapters will examine the neuroscience behind this transformation. As we'll see, music is not just an art; it's also a tool for lifelong learning and growth.

Antiquity Of Music

How old is music, really? To find out, we turn to the archaeological record, and what it shows is astonishing. The oldest known musical instrument is a flute carved from the thighbone of a young cave bear. It's about 60,000 years old. Found in the Divje Babe cave in Slovenia, this ancient flute has carefully drilled holes and a design that suggests it was intentionally tuned. Experts believe Neanderthals made it since they thrived in those regions before modern humans arrived in Europe.

However, the complexity of this instrument suggests something even older. To craft such a flute, someone must have already mastered playing and designing similar instruments. Musical knowledge must have existed long before this flute was made.

Humans never needed instruments to make music in the first place, though. The body itself is a musical instrument. We can clap, stomp, or use our voices. Chanting, humming, or singing—especially in rituals—likely came long before flutes and drums. Vocal music may well be the earliest form of human musical expression.

There is even anatomical evidence supporting this ancient origin. Fossils of Homo *heidelbergensis*, a probable ancestor of both modern humans and Neanderthals, include a hyoid bone—crucial for vocalization—that closely resembles ours in shape and spectral qualities. This implies that the ability to produce a wide range of vocal sounds, maybe a few octaves, may date back as far as 500,000 years. Of course, hearing is part of the musical chain; research on audiology shows that Neanderthal hearing ability is comparable to that of modern humans (Chapter 5).

In short, music isn't a recent cultural invention. It predates civilization, agriculture, and even modern humans, i.e., Homo *sapiens*. It's woven into the deep fabric of human evolution. The next question is: why has music been a part of our lives for so long? Is it essential for survival, or just a beautiful byproduct of our complex brains? That's precisely what we should ask next.

Do Humans Need Music?

Considering music's deep roots and presence in every culture, one might ask: Do we truly need music, or do we simply enjoy it immensely? While some believe that we gain benefits from music and that it offers certain evolutionary advantages, others contend that it has no real value for humanity and is merely a means to convey messages in addition to languages.

The Nay camp believes that music has no value for humans and their evolutionary history. Population anthropologist Steven Pinker, an American philosopher and psychologist influenced by William James, argued that music is unimportant to modern humans. He famously claimed that music is a spandrel, a useless evolutionary by-product of another trait, specifically language. In his 1997 book, "How the Mind Works," Pinker referred to music as "auditory cheesecake" and stated that it serves no biological function, being merely an invention that people enjoy, similar to the taste of cheesecake.

The most prominent 19th-century scientist and thinker, Charles Darwin, famously observed: "*As neither the enjoyment nor the capacity to produce musical notes are faculties of any use to man in relation to his daily habits of life, they must be ranked among the most mysterious with which he is endowed.*" In other words, from a practical standpoint, he acknowledged that our musicality is not useful in everyday life, yet we possess this ability. It is undoubtedly an evolutionary and philo-sophical puzzle that he could not easily resolve.

However, the Nay camp may have overlooked that music has always been an activity involving many individuals, as it was in our early history. Early humans lived in groups smaller than 150 individuals, according to R. I. M. Dunbar, and these groups can be seen as evolu-tionary units. It might be possible that most members of the group participated in musical activities for some intuitive reason. Wouldn't that create a group-specific character that could either provide or hinder the group's evolutionary advantages?

The Aye camp holds a different view, based on the idea that music is everywhere in our lives and history. It must have served some biological and evolutionary functions. Moreover, there is strong neuroscientific evidence showing that our brains are built to handle and process music, whether we are aware of it or not. Why would we develop this ability if it doesn't provide any evolutionary benefits, even if it could be subordinate to language? Yet, there is no direct evidence that definitively links music to evolution; any such connection remains speculative and unproven.

Philosophers like Adam Smith, David Hume, and Arthur Schopenhauer argued that the desire to synchronize one's rhythms with others, such as in music, is a basic drive for a sense of belonging among members. As a result, this feeling promotes and reinforces empathy, which leads to altruistic actions and stronger, more cohesive groups. Would this be seen as a survival trait from an evolutionary perspective?

Strong disagreements on this issue persist, with no final judgment in sight; however, the debates and ongoing discussions continue to deepen our understanding of music's importance in practical, spiritual, and philosophical contexts. Perhaps a better scientific grasp of how humans naturally possess musicality can help clarify the debate a bit.

The questions about music listed above are not meant to be comprehensive. However, the answers to them, whether simple or complex, tend to lean toward science for solutions. Still, the predominant scientific studies remain phenomenological in nature. Most efforts to answer are adequate, but they fall short of connecting the various aspects of music under a single principle to which we can always refer. The goal of this book is to address this underlying unifying principle while providing an overview of the sciences with the necessary details.

Sciences of Music

Summary of Scientific Musical Study

The scientific interest in music goes back more than two thousand years. One of the earliest thinkers to study it seriously was Pythagoras, the Greek mathematician and philosopher from the 6th century BCE. While best known for his geometric theorem, Pythagoras was also captivated by the connection between musical tones and mathematical ratios. Using the limited scientific tools available in his time, he examined how the length of a string influenced pitch and introduced early concepts of harmony and consonance.

However, the serious scientific study of music didn't begin until the 19th century. In 1877, physicist Hermann von Helmholtz published *On the Sensations of Tone*, a foundational work that explored how sound waves interact with the ear and brain. He established the foundation for the study of auditory perception and musical acoustics.

At the start of the 20th century, research on music was divided into two main areas. One group, including physicists and engineers, examined sound as a physical phenomenon, known as acoustics. The other group, made up of psychologists and musicologists, studied the emotional and behavioral effects of music. These efforts often remained separate, with little interaction. The subjective and emotional power of music was usually seen as too mysterious for rigorous scientific study.

That started to change with the rise of modern neuroscience and cognitive psychology. As brain and neural science advanced, our ability to study music in a more integrated way also improved. Since the 1960s, music has become an important subject for researchers from various fields, including neuroscience, psychology, linguistics, and physics. Advanced brain imaging tools like fMRI and EEG have shown that specific brain areas light up stronger than usual in response to different musical elements, such as pitch, rhythm, melody, and lyrics. These findings have helped show that music isn't just heard but is also processed and experienced through extensive networks in the brain.

Some of the pioneers in this field include Aniruddh Patel, Daniel Levitin, Robert Zatorre, and Oliver Sacks. Levitin's book This Is Your Brain on Music brought many of these ideas to the public. Zatorre's research revealed details of how music interacts with the brain's reward systems. And Oliver Sacks, through his fascinating case studies, demonstrated how musical abilities can persist even in the face of brain damage, highlighting music's unique relationship with memory, identity, and emotion.

Even physicists are interested. In fact, research work on how we perceive musical consonance has been published in some of the top physics journals, including *Physical Review Letters*. These studies treat music perception as important as discoveries in fundamental physics. There is one particular work, published in the journal after thorough peer review, that holds academic weight comparable to the discovery of the Higgs boson.

As a sidebar, the significance of the Higgs boson lies in the existence of a particle that makes up part of what we observe in the universe. The idea that research on music in physics has the same importance as the Higgs boson discovery highlights how music is respected as a scientific field. Incidentally, I also had a few articles published in the same journal when I had the privilege to collaborate with a twice-nominated Nobel physicist (who did not win, but whose scientific achievements are undeniable) in the early 1980s.

The result is a growing field that treats music as a legitimate subject of scientific study. From vibrations in the air to dopamine in the brain, music is now being studied across a broad spectrum, including physics, psychology, and neuroscience. It is a genuine academic discipline.

However, even with these advancements, we are still far from fully understanding how and why we are so drawn to music. Much of what we know remains phenomenological: we can observe what happens when people hear music, but the deeper unifying principles —those that explain why music affects us so powerfully as a human species—are still emerging.

This book suggests that these deeper connections exist and are based on physical laws that govern not just music, but all of nature.

Accomplishments in the Neuroscience of Music

In recent decades, neuroscience has made significant progress in understanding how music affects the brain and its activities. What once seemed abstract—our emotional response to music—has become more measurable and explainable through scientific study.

One major breakthrough has been the mapping of specific brain regions to musical functions. Listening to or creating music activates a network of areas: the auditory cortex processes sound, the motor cortex tracks rhythm and movement, the limbic system handles emotion, and the prefrontal cortex supports attention and memory. This reveals that music is not processed in isolation. It's a full-brain experience, involving emotion, cognition, memory, and movement all at once.

Daniel Levitin's book, "*This Is Your Brain on Music*," helped expand this insight to a wider audience. While it focuses more on trained musicians, researchers like Robert Zatorre have demonstrated that these effects apply to everyone. Musical perception activates neural reward systems in both musicians and non-musicians, indicating that musical enjoyment is a natural part of the human brain.

Zatorre and Anne Blood were among the first to show that music can activate the same brain regions as food, sex, and social bonding can. When we listen to emotionally powerful music, our brains release dopamine, a neurotransmitter linked to pleasure and reward. These responses are strongest in areas like the ventral striatum, amygdala, and prefrontal cortex. In other words, music can create physical sensations, such as chills, goosebumps, and tears, that are chemically and neurologically real.

Rhythm also has deep roots in brain function. The cerebellum, motor cortex, and basal ganglia all participate in tracking and predicting beats. This has practical implications. In clinical settings, rhythmic music has been shown to help patients with Parkinson's disease or stroke improve coordination and walking. Research by Michael Thaut and others has demonstrated that rhythmic auditory stimulation can retrain motor systems and restore movement through a process called entrainment.

Music also plays a crucial role in memory. Studies have shown that music can improve recall in individuals with neurological impairments, such as Alzheimer's disease. A 2007 study by Petr Janata found that familiar songs triggered autobiographical memories in Alzheimer's patients—memories that were otherwise difficult to access. Music activates the hippocampus and other regions linked to long-term memory, helping to maintain a sense of identity even when other systems fail.

Language and music share many neural pathways. Both depend on timing, tone, and structure, and both activate regions like Broca's and Wernicke's areas, which handle speech production and understanding. Musical training has been shown to enhance language processing, and the reverse is also true.

Neuroplasticity—the brain's ability to adapt—also plays a crucial role. Musicians often show enhanced connectivity between auditory, motor, and cognitive regions. Even brief musical exposure can physically modify the brain. A 2005 study by Gottfried Schlaug found that children who received music training showed increased gray matter in motor and auditory areas. These changes indicate that music not only affects the brain but also helps develop it.

Together, these achievements have transformed music neuroscience from a niche interest into a recognized and expanding field. Music is now acknowledged not only as a source of enjoyment and art but also as a scientifically meaningful tool for understanding how the brain functions—and how it recovers.

Beyond Music Phenomenology

Taking a step back, we can see a broader picture emerging from decades of neuroscience research into music. The discoveries are exciting, and the achievement is commendable. We know where music is processed in the brain, how it can regulate emotions, and how it activates reward systems. However, the root cause of our deep obsession with music remains less fully understood.

Think about what actually happens when we hear music. Sound waves travel through the air, influenced by acoustics—a field governed by physics. These waves hit our ears, where they are transformed into mechanical and then chemical signals. These signals are sent to different parts of the brain, activating auditory pathways and eventually linking to memory and emotion. In the end, they create sensations that feel deeply personal and meaningful.

Each step in this chain involves a different scientific discipline: acoustics, physiology, neurobiology, and psychology. And yet, the entire process, from vibrations in the air to chills down your spine, is seamless to the listener. It feels like magic.

Is there a principle that connects all these processes? Despite our advances, the sciences of music still often work in isolation. Each research area — hearing, perception, memory, and emotion — has developed into its own highly specialized field. But one can gain deep insight into the subject at the expense of the broader view, a common analogy in photography. This is the tradeoff of academic depth: we gain precision but often lose perspective.

What we need is a way to zoom out and see the entire landscape of music's effects through a unifying lens. This book suggests that such a unifying principle exists. It argues that the scientific processes involved in music—from sound waves to emotional reactions—are all governed by a deeper law: a physical principle rooted in the brain's effort to predict, regulate, and minimize energy expenditure. Like gravity in physics, this principle could provide a simple explanation for a complex system.

The brain, like all physical systems, operates according to the laws of thermodynamics. It continuously seeks efficient pathways, minimizing uncertainty and conserving energy. In music, this appears as a prediction engine that responds strongly to patterns, surprises, and variations in sound. These responses trigger emotional reactions and the rewarding feelings we get from music.

This approach doesn't reduce music to cold mechanics. On the contrary, it helps us understand why music feels the way it does. It provides a framework connecting biology and emotion, physics and psychology, as well as acoustics and aesthetics.

In this book, we will explore a scientific and engineering model that seeks to explain music as a system governed by feedback and feed-forward control mechanisms—structures that predict, respond, and adapt in ways consistent with the second law of thermodynamics. This model not only provides a way to explain musical behavior but also connects the sciences of music into a coherent whole.

Physics Perspective on Cognition and Learning

To understand how music moves us, we need to ask a deeper question: what underlying process governs our response to sound? Across different senses—sight, sound, touch, taste, and smell—our brains rely on one fundamental strategy: statistical learning and predictive coding.

This process lets the brain absorb sensory information, compare it with past experiences, and predict what will happen next. When those predictions come true, the system stabilizes. When they are defied in just the right way, such as with a surprising chord change or a rhythmic twist, the brain lights up with reward. That's often where pleasure happens.

Predictive coding is not unique to music. It's crucial for how we learn language, interpret visual scenes, and navigate social interactions. However, music provides an especially elegant and emotionally rich environment for this process. That's why statistical learning and predictive coding (SLPC) are key to understanding musical experience at every level—from raw sensory input to emotional excitement.

But what drives this learning and prediction process? Here we present an intriguing idea: that the second law of thermodynamics, a fundamental principle in physics, operates behind the scenes of our musical minds. This law states that physical systems naturally move toward states of lower energy. It explains why heat flows from hot to cold and why stars burn out. Surprisingly, this same principle may also apply to neural activity and music.

When the brain processes music, it constantly updates its predictions to minimize uncertainty and optimize its internal state. This helps conserve energy. If the sounds it receives match its expectations, the brain can relax—no adjustments are needed. If the music defies expectations, the brain responds by updating its model while reacting emotionally or physiologically. These changes are the brain's way of trying to restore balance, or in thermodynamic terms, to a "lower energy" state.

This feedback/feedforward loop—adjusting predictions based on incoming signals and restoring balance—is central to how we perceive music. Since it occurs across different brain regions, it connects everything from auditory input to motor coordination, memory recall, and emotional response.

This idea will be central to the rest of the book. We'll explore how the second law of thermodynamics can serve as a unifying framework for understanding music—not just as a series of enjoyable sounds, but as a profound biological and physical process rooted in how our brains aim to function efficiently.

In this perspective, music becomes more than just art. It acts as a reflection of how the brain—and maybe the entire human system—balances chaos and order, energy and emotion, prediction and surprise.

What This Book Brings to the Table

When I first began exploring music writing through the lens of physics, many of my friends were skeptical. Shouldn't someone who writes about music know how to play an instrument or compose a symphony? I don't have a good answer. I'm not a musician in the traditional sense. I can't perform or write scores. But I am—like many of us—a passionate listener.

Music moves me. It excites me, surprises me, and sometimes gives me chills. I feel a deep connection to certain pieces, even though I can't reproduce them or explain why they resonate so strongly. What does it mean to love music but not be able to create it? Where does that emotional power come from?

These questions inspired me to view music not as an artist or critic, but as a scientist. My background is in physics and engineering. Over the years, I've published technical papers in these fields and spent much of my career developing the foundational technologies behind the internet—specifically, optical communication systems. However,

I've also ventured outside my domain to investigate broader scientific questions about human origins and evolution.

In 2021, I published a popular science book titled *From Where We Came: A Physicist's Perspective on Human Origin, Adaptation, Proliferation, and Development.* In that book, I explored human evolution from a molecular perspective, showing how a single physical principle, random mutation and its mathematical formulation, combined with thermodynamics, can help explain the incredible diversity of our species. That book was later published in China and is being translated into other languages.

Now, I want to bring that same perspective to music. This book offers a new way to see music—one that is scientific, but also accessible. By finding a unifying physical principle behind our musical experiences, I aim to create a framework that is intellectually rigorous but easy to understand. You don't need to be a musician or a scientist to read this book. You just need to be curious.

Music belongs to everyone. If we can understand why it affects us so profoundly through science across cultures and generations, we can better appreciate not just music but also the remarkable brains that created it.

Holistic Approach to Musicality

To truly understand music, we need more than just isolated facts. We need a broader framework—one that links the physics of sound, the biology of hearing, the neuroscience of perception, and the psychology of emotion. Music doesn't exist in parts. It happens as a process, and that process crosses disciplinary boundaries.

This book proposes and advocates a holistic scientific approach to music. At the heart of this approach is a unifying idea: that the human experience of music, from generation to perception to reward, can be understood through a single underlying principle.

That principle is rooted in the second law of thermodynamics: nature's tendency toward states of lowest energy.

To establish a connection between thermodynamics and music, I draw not only on neuroscience and psychology but also on engineering, specifically from a familiar engineering system called a control system, which can be operated in a feedback, feedforward, or both fashion. Control systems are used in everything from thermostats to autonomous vehicles. They operate by comparing a desired state with the current state, making adjustments as needed, and working toward equilibrium. Our brains undergo a similar process when processing music.

One can think of the entire musical experience as a series of transformations. Music starts as a physical disturbance, vibrations in the air. Instruments or voices shape these vibrations and send them to our ears. There, they are converted from mechanical signals into neural impulses and transmitted to the auditory cortex. The brain then decodes these impulses into tones, rhythms, and melodies. Next, cognition and memory assign meaning, which triggers emotional reactions and a sense of reward in our innate reward system.

Each step is governed by its own rules. The physics of sound doesn't explain the psychology of pleasure. But what connects these steps is the brain's effort to stabilize, predict, and adjust—constantly seeking a lower-energy, more efficient state. This is where feedback and feedforward control loops come into play. Just like an engineer adjusting a system, the brain fine-tunes its response to music through nested loops of prediction, correction, and reward.

To make this train of events easier to understand, I introduce a concept I call musical phased-naturalism. This framework views each stage of the musical experience as a "phase" governed by specific scientific laws. The shift from sound creation to hearing, from hearing to perception, and from perception to emotional reaction involves changes across domains: acoustics, physiology, cognition, and psychology.

Each domain uses a different "language." Acoustics depends on wave equations. Neuroscience utilizes electrochemical signals. Psychology communicates in terms of emotion and meaning. However, across all these levels, the system follows the same principle: it seeks balance through feedback/feedforward and prediction, guided by fundamental physical laws.

This perspective doesn't simplify music. Instead, it breaks down the whole music experience into manageable, semi-independent entities for in-depth research on each, but they can be reassembled for the holistic picture. Music is a system, an ordered one, created not just by any single mind, but also through the natural functions of the brain and the laws of physics that influence it.

<u>Musical Phased-Naturalism</u>

To understand music scientifically, we can view it as progressing through a series of interconnected "phases." Each phase belongs to a different domain, such as acoustics, physiology, neuroscience, or psychology, and each follows its own rules. Putting them all together, it forms the musical phased-naturalism framework.

This idea of "divide and conquer" isn't new. In physics, we often analyze complex systems in layers. We study atoms, molecules, solids, and galaxies, each considered its own branch of science. Yet beneath these fields is a shared principle: all systems naturally tend toward lower-energy states. This principle, rooted in the second law of thermodynamics, applies across all scales, from quantum particles to the entire universe. Music can be viewed similarly. We can visualize the concept of musical phase-naturalism as shown graphically in Figure 1.1, which illustrates the five phases.

Musical Phased-Naturalism

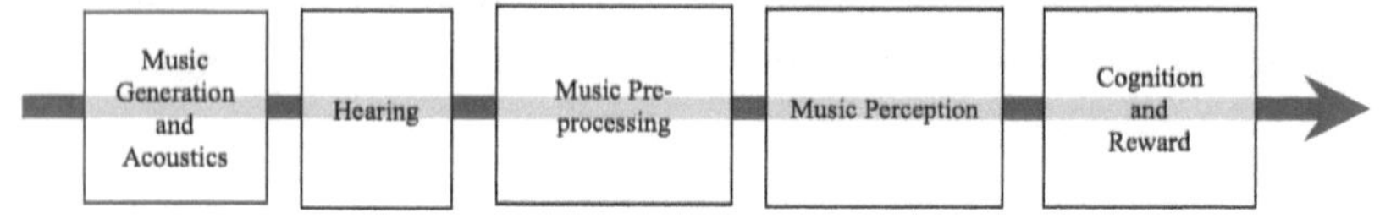

Figure 1.1. The graphical representation of the phased-naturalism formulation. The five quasi-independent phases that thread through the chain of musical process. They are music generation and acoustics, hearing, music pre-processing, perception, and finally, cognition and reward.

The first phase is sound generation. This involves the physical process of creating musical tones, whether through voice, instrument, or electronic technology. Sound consists of vibrations, specifically pressure waves in the air that can be described by frequency, amplitude, and waveform. This falls under the field of acoustics. Although included in the phased naturalism, it involves the feedback/feedforward loop from emotion back to the generation, it is less directly connected to neuroscience. We will not expand on this phase.

The second phase is hearing. When those vibrations reach our ears, they are mechanically processed by the eardrum and inner ear, then converted into neural signals. This phase involves a transformation from physical to electrical signals, governed by auditory physiology.

The third phase is preprocessing. Here, electrical signals are mapped onto the auditory cortex without yet assigning emotional meaning. This phase acts like a sensory relay, organizing sounds by frequency and timing but not interpreting them. It maintains most of the auditory information. The original auditory signals are gradually transformed into something between raw data and processed information as they ascend through the neurological organization. Therefore, it resides between physiology and neuroscience.

The fourth phase is perception. At this stage, the brain begins processing incoming musical information beyond its purely "auditory" qualities, extracting melody, rhythm, tempo, harmony, and structure. It compares these features to existing memories and expectations. It is considered a distinct phase because it involves a different depth of the music, with signaling now focused on passing analyzed messages rather than raw data.

The subjects and activities in this phase, along with those in the next phase, primarily fall within the field of neuroscience. Mentioning neuroscience may seem challenging for general readers. Some background in neuroscience could be helpful, especially when discussing the functions and locations of brain regions using their specific terminology and methods. However, prior knowledge is not required because the terminology is less important than the systematic and logical conclusions drawn from many well-designed experiments, supported by modern diagnostic tools. The complex terminology and names will soon become familiar and understandable as we progress. Compared to other well-established sciences, this relatively new field can be understood logically as long as one develops a basic understanding of brain structure and its functions.

The fifth and final phase focuses on neural responses and rewards. The brain's emotional and reward systems now evaluate the perceived music. When the analyzed music matches memories and expectations, the reward system reacts with pleasure, mediated by neurotransmitters like dopamine. This phase involves memory, emotion, and social psychology.

What makes this model helpful is not just that it organizes musical experiences into stages. It also illustrates how these stages create a feedback and feedforward system. Each phase influences the next but also loops back in a feedback or feedforward manner (the concepts of feedback and feedforward will be explained in Chapter 4). For example, if a rhythm surprises us, our perception shifts, which then affects our expectations, leading to emotional responses or physical movements. It is a chain of processes feeding backwards and forwards through the brain structure.

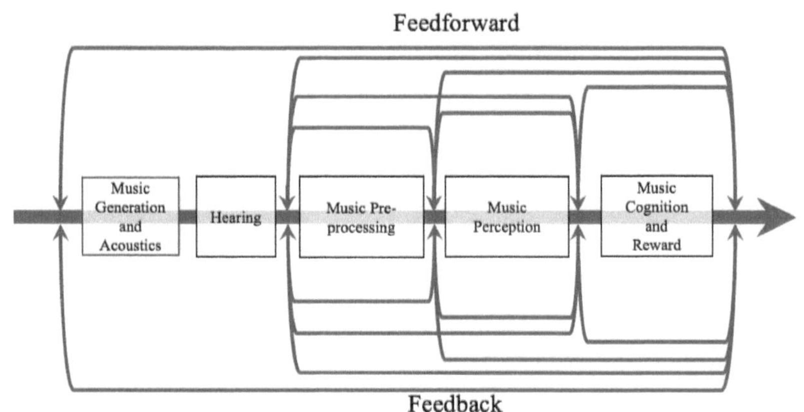

Figure 1.2. *The control system concept of the phased-naturalism. The multiple feedback and feedforward loops constitute the nested control system. They also signify the interactive nature of these phases.*

These nested feedback and feedforward loops, as shown in Figure 1.2, help explain how we learn music, anticipate changes, experience tension and resolution, and ultimately derive emotional rewards. They operate within the brain just as they do in engineered control systems, all while obeying thermodynamic principles that aim to conserve energy and maintain their lowest energy states.

In summary, musical phased-naturalism provides a framework for understanding how music moves from the external world into our inner experience, while presenting detailed scientific information in an organized way.

A Unifying Physics Law

At the core of this book is a new idea: that the whole process of engaging with music, from sound creation to emotional reaction, is governed by a single physical law. That law is the second law of thermodynamics, which states that all physical systems will naturally tend to move toward the lowest energy and highest entropy unless influenced by external forces.

This principle is well-known in physics. Stars burn out; heat transfers from hot objects to cold ones; molecules reach equilibrium in chemical reactions. However, this same drive toward balance and efficiency also governs behaviors in living systems and the human brain.

In the realm of music, this law demonstrates how the brain constantly seeks predictability and stability. Every aspect of the musical experience—hearing, perception, cognition, and emotional responses—functions as a system trying to remain in or return to its lowest energy state. When we listen to music, our brains predict what will happen next. If the prediction is correct, energy is conserved. If not, the system must work harder to regain balance, which is the lowest energy state.

This feedback/feedforward process is similar to what we see in engineering control systems. When a thermostat detects that a room is too cold, it sends a signal to activate the heater. Once the desired temperature is reached, it turns off, maintaining balance. The brain works in a similar way when responding to musical stimuli. If music follows a predictable pattern, such as a familiar melody or a steady rhythm, the brain feels comfortable and can simply coast. However, if music introduces a surprise, such as a sudden modulation, a change in tempo, or a harmonic twist, the brain becomes alert. It responds to novelty, recalculates expectations, and rewards itself when the tension is resolved. This entire loop reflects the brain's effort to manage its internal state with minimal energy expended.

Throughout the rest of this book, we will examine how this control system and its underlying physics principle work at each phase of musical processing, from sound waves in the air to dopamine releases in the brain. We will explore how musical expectations are formed, how they are disrupted, and how the brain responds through sequences of neural feedback. In this way, music provides not only emotional depth but also a window into the laws that govern our biology and behavior.

Book Outline

For the rest of the book, I invite you to explore music from this all-encompassing perspective, considering the simplest biological system, the emerging categories of segmented science, and their corresponding roles within phased naturalism. The chapters are organized slightly differently from the phased naturalism sequence shown in Figure 1.1. Instead, they will follow a logical progression for clarity.

This first chapter explores a range of relevant musical topics, beginning with the basic question of why we find music so fascinating. It also introduces, for the first time, the connection between each scientific musical step and a unifying natural principle that supports them. The behaviors associated with each step resemble a control system designed to balance the core stimuli of music with perceptual and cognitive processes, which then lead to emotional and psychological responses in individuals and groups.

Chapter 2 provides a detailed overview of our deep connection with music, examining musical activities, culture, and history. It summarizes our relationship with music through various phenomena without delving deeply into music itself, functioning as a phenomenological overview. It becomes evident that nearly every aspect of music is woven into our lives and influences us across different historical periods and locations. Ultimately, this chapter reflects on our extensive relationships with music by considering past events and current influences in culture, politics, and all areas of life. These facts reinforce the idea of our innate musicality and human obsession. This relationship touches many parts of our past and future lives, making the chapter's narration necessarily broad, comprehensive, and detailed as needed.

Chapter 3 examines the universality of musical scales. It emphasizes the similarities found in different cultures around the world. This similarity stems from a shared neurological basis across our species, including modern humans and Neanderthals (yes, we are the same species, even if we belong to different subspecies). It will be demonstrated that many cultures have embraced this commonality because of our biological tendencies, which cause us to converge on "identical" scales when viewed through a mathematical lens. Additionally, this chapter provides a preview of the neuroscientific evidence supporting the recognition of octaves, which underpins scales across all cultures. The detailed explanation of this neuroscientific foundation is the main focus of Chapters 7 and 8.

After providing a brief historical overview of the scientific and neuroscientific aspects of music in Chapters 1, 2, and 3, Chapter 4 offers a crucial scientific overview of the phased-naturalism framework that underpins every aspect of the SLPC and its guiding principles, specifically, the second law of thermodynamics. This chapter presents this framework from two perspectives. The first perspective is a universally applicable foundation of the second law of thermodynamics, which is inherently integrated into all branches of physics and science. The second perspective explains how this principle is modeled through an engineering concept of a control system.

Chapter 5 explores how we hear music from the perspectives of biology and physiology. It thoroughly examines the fundamental physiology involved in transforming air into neural impulses. The chapter explicitly compares the timing and location of hearing with the functions of sound waveforms and spectra. It also explains how our inner ear converts acoustic disturbances into electrical neural signals that precisely reflect changes in air pressure. This process marks the initial stage of our acoustic physiology. Additionally, it connects the waveform and spectral analysis performed by the inner ear to physics and engineering concepts for better comparison and understanding.

Both the anatomical and functional connections involved in how music is processed in the auditory cortex will be explained in Chapter 6. Our understanding of auditory pathways benefits from detailed neuroscientific research on language, utilizing advancements in scientific tools that can monitor brain activity in various ways after exposure to sound and language stimuli. This chapter examines how musical signals travel through hierarchically organized brain regions from the ears to the tonal processing area of the auditory cortex. Feedback and feedforward mechanisms are present in multiple pathways and are clearly demonstrated here. Specifically, this chapter discusses the gateway that transforms physiological processes into complex neural computations, leading to perceptions.

Chapter 7 explains that musical perception is how the brain responds to hearing music. The auditory nerves and impulses have no meaning until this stage. Perception involves feelings, whether pleasant, sad, or uncomfortable. It originates from the ventral and dorsal areas of the auditory cortex, continuing the process of perception. At this point, auditory impulses in the neurons no longer directly represent the acoustic features; instead, they are further interpreted by the ventral and dorsal regions to uncover meaning from the raw data. The ventral pathway (ventral stream) processes the tones, melody, and harmony of music within the audio frequency spectrum. The dorsal pathway (dorsal stream) interprets music's variations over time, including pace, beats, and rhythms.

Both streams are bidirectional, working together with memory in the hippocampus. They facilitate rapid upward signal flow and slower downward signal flow. However, if the stimuli match the existing memory in those areas, this part of the system remains unchanged, conserving energy since it is already in the lowest energy state. If the stimuli differ, the system sends feedback indicating a different perception, suggesting that memory needs updating.

Chapter 8 explains how the perception created by the auditory cortex and other regions interacts with our reward system both anatomically and functionally. The pleasure and signals of enjoyment from the reward system triggered by music are similar to those produced by other sensory functions. Music's acoustic and neurophysiological nature has transformed into messages that are understandable by the reward system. The ways musical signals eventually reach our reward system are described, sometimes directly and other times indirectly.

Musical signaling serves as an indicator of the expectations and anticipations that create our pleasure, expressed through sensations like chills or emotions such as sweating and electrical conductance of the skin. Eventually, the emotions and psychological responses initially triggered by the musical stimulus are linked to feedback and feedforward effects observed in the release of neurotransmitters, such as dopamine and other chemicals. This results in musical sensations and psychological experiences that manifest as a range of emotions, including pleasure and sadness.

This chapter highlights that, although the pleasure and sadness discussed earlier are personal feelings, the desire to share experiences with others in both public and private settings is undeniable. It provides convincing evidence that music triggers a primal and emotional reward that brings us together. This is clearly seen in people trying to synchronize their musical movements and emotions within groups, leading to the phenomenon of music group psychology. It can create strong feelings of community, ideology, and shared visions.

Chapter 9 explores the intuitive judgment of music's essential elements: harmony and tones. It is not a separate aspect of the event sequence from the phased-naturalism formulism. It is demonstrated that the ability to judge harmony can occur early in our neural hierarchy, before any complex neural processing takes place. This finding further supports the idea that humans have innate musicality. The chapter also shows that studying consonance and dissonance physically can help resolve the ongoing debate about consonance in neuroscience. Musical research indeed requires collaboration among various academic disciplines that complement each other for a deeper understanding.

Summary

This book briefly explores the individual sciences and neuroscience within a phased naturalism framework, based on a fundamental physics principle: the second law of thermodynamics. In the final chapter, we look at the still-evolving scientific understanding of music, hoping that ongoing research will keep satisfying our curiosity as we aim to appreciate it in all its wonder.

Most importantly, the essential role of the second law of thermodynamics should be recognized, as it clarifies some aspects of the science of music that no other field can offer. We will highlight this perspective whenever appropriate throughout this book.

It also raises questions about whether our scientific and neuroscientific views on music diminish our appreciation for its artistic qualities and enjoyment. Does this mean the artistic side of music becomes insignificant when we focus on science? The answer is a definite "no" because the science and neuroscience of music are still far from complete; whether we continue our scientific pursuit has not changed our love of music through history, let alone the continuing study. More importantly, even if we were to achieve a full understanding of the fundamental science, it would not lessen the natural biological pleasure we get from listening to and performing music. So, while we continue exploring the neuroscience, physics, and other scientific aspects of our passion for music, its powerful emotional impact on us remains the same.

Let's fulfill our desire to be rewarded, captivated, and intoxicated by participating in musical activities, regardless of the scientific principles behind the music.

Bibliography

1. "Early Neanderthal Constructions deep in Bruniquel Cave in southwestern France," Jaubert J, Verheyden S, Genty D, Soulier M, Cheng H, Blamart D, Burlet C, Camus H, Delaby S, Deldicque D, Edwards RL, Ferrier C, Lacrampe-Cuyaubère F, Lévêque F, Maksud F, Mora P, Muth X, Régnier É, Rouzaud JN, Santos F. *Nature* **534**, 111-4, 2016.

2. "From where we came, a physicist's perspective on human origin, adaptation, proliferation, and development," C. Y. Kelly, *CYK Publishing House*, 2021.

3. "Brain Indices of Music Processing: 'Nonmusicians' Are Musical," Stefan Koelsch, Tomas Gunter, and Angela Friederici, *J. Cognitive Neuroscience*, **12:3**, 520-541, 2000.

4. "Neuroscientific Insights for Improved Outcomes in Music-based Interventions," Psyche Loui, *Music and Science*, **3**, 1-9, 2020.

5. "Timbral effects on consonance disentangle psychoacoustic mechanisms and suggest perceptual origins for musical scales," R. Marjieh, P. Harrison, H. Lee, F. Deligiannakie, N. Jacoby, *Nature Communications*, **15**, 1482, 2024.

6. "The Big Picture: On the Origins of Life, Meaning, and the Universe Itself," Sean Carroll, *Dutton Publisher*, 2016.

7. "Our Modern Skulls House a Stone Age Brain: An Overview and Annotated Bibliography of Evolutionary Psychology," Tony. Stankus, *An Overview and Annotated Bibliography of Evolutionary Psychology, Part I, Behavioral & Social Sciences Librarian*, **30**(3), 119-141(2011). DOI: 10.1080/01639269.2011.591276

8. "New flutes document the earliest musical tradition in southwestern Germany," Conard Nicholas J., Malina, Maria, Münzel, and Susanne C, *Nature* **460**, 737–40, 2009.

9. https://www.classicfm.com/discover-music/latest/best-classical-flash-mobs/

10. "Oldest playable musical instruments found at Jiahu early Neolithic site in China," J. Zhang, G. Harbottle, C. Wang, and Z. Kong, *Nature*, **401,** 366-368, 1999.

11. "Brain indices of music processing: 'nonmusicians' are musical," S. Koelsch, T. Gunter, A. Friederici, and E. Schroger, *J. Cognitive Neuroscience*, **12**(3), 520-541, 2000. doi: 10.1162/089892900562183

12. "From Perception to Pleasure: The Neuroscience of Music and Why We Love It," R. Zatorre, *Oxford University Press*, 2023.

13. "This Is Your Brain on Music: The Science of a Human Obsession," Levitin, D. J. Dutton/Penguin (2006).

14. "Music, Language, and the Brain," 2008 Patel, A. D.

15. "Intensely pleasurable responses to music correlate with activity in brain regions involved in reward and emotion," Blood, A. J., & Zatorre, R. J., *Proceedings of the National Academy of Sciences,* **98**(20), 11818–11823 (2001). https://doi.org/10.1073/pnas.191355898

16. "Neural basis of rhythmic timing networks in the human brain," Thaut, M. H., & Davis, W. B. *Journal of Cognitive Neuroscience*, aljh, (1997).

17. "Characterizing the music-evoked autobiographical recall (MEAR) phenomenon," Janata, P., Tomic, S. T., & Rakowski, S. K. *Psychological Science*, (2007).
18. "The effects of musical training on the brain and cognition," Schlaug, G., et al.. *Annals of the New York Academy of Sciences*, (2005).
19. "Neocortex size as a constraint on group size in primates," R. I. M. Dunbar, *Journal of Human Evolution,* **22**(6), 469–493 (1992). https://doi.org/10.1016/0047-2484(92)90081-J
20. "Synchrony and the social tuning of compassion," P. Valdesolo, D. DeSteno, *Emotion,* **11**(2), 262–266, (2011).

Chapter 2

The Causal Connection

Musicality and Obsession

Before exploring the sciences of musical phased-naturalism, this chapter aims to establish a fundamental premise: humanity is truly deeply obsessed with music. If true, this widespread fascination cannot be coincidental. There must be something innate, familiar to us, and part of being human. We assert that this is our ability to engage with and be moved by music, that is, musicality.

But what exactly is musicality anyway? While definitions vary, this book adopts a working definition: musicality is our innate ability to understand, respond to, and engage in musical activities. Musicality isn't just about technical skill. It reflects emotional sensitivity and expressiveness. It includes timing, coordination, and the ability to interpret musical cues through listening, performing, or composing. While exposure, experience, and training can enhance it, musicality originates from our neural structure; it's hardwired into us. This might explain why we can quickly adapt to different musical styles, create new sounds, and feel music deeply. This natural instinct allows us to learn and appreciate music, even without formal training.

What does it mean to say we're obsessed with music? This isn't about individual taste; it's about music's constant presence in human life. Music permeates all cultures, eras, and social structures. Even Neanderthals, our ancient extinct relatives, likely shared in this passion.

This chapter explores how the connection between passion and musicality emerges from various situational factors. We aim to show, through human behavior, that cognitive neuroscientist Stefan Koelsch's claim is correct: even untrained listeners, often called "non-musicians," demonstrate a strong understanding of music not through words but through "feeling." We will examine the practical evidence of this idea in Part I.

In Part II, we'll examine this relationship further from the perspective of the "effects part of causality." It will be clear that the effect is seen throughout human history and appears in every aspect of our lives.

The musicality and obsession, as defined above, are present in almost every human endeavor. In part III, we examine these presences within the broader context of "music's many intersections" with various aspects of our lives. By the end of the chapter, you'll see how the cause (innate musicality) and the effect (obsession) are two sides of the same coin. Our ultimate goal is to demonstrate that our connection with music is so profound that it sparks a curiosity that becomes an infatuation.

Part I: Musicality

<u>Personal Musical Journey</u>

I'm not a trained musician. Still, I consider myself a deeply musical person. I remember hearing Aaron Copland's Hoedown about fifty years ago; it conjured a vivid image of people, children and adults alike, dancing at a country fair, tambourines jingling, violins soaring. The uplifting rhythms and the bright melody took hold of me. I clapped, tapped, hummed, and swayed without thinking. Even now, the tune lingers. It had no lyrics, just joy. And that joy left a lasting impression.

That's how music works for me: I gravitate toward anything that feels good, soothing, energizing, and nostalgic. I love music from all genres: classical, rock and roll, R&B, musicals, Western, and Eastern, as well as 40s to 80s pop. If it moves me, I'm in.

Much of my musical taste is rooted in American culture. I admire Gershwin, Glenn Miller, Elvis, The Beach Boys (not boys anymore), and pop icons from the British Invasion, The Beatles, Elton John, and Tom Jones. I also appreciate Chinese symphonic works, such as "The Butterfly Lovers" and "The Yellow River Concerto," which blend Western structure and instrumentation with Eastern melodies and emotion.

Film music holds a special place in my heart. John Williams and Hans Zimmer's compositions breathe life into cinema. Whether it's Superman's iconic theme, The Lion King's emotional resonance, or Gladiator's somber tunes, these scores elevate storytelling. Why I get goosebumps and chills following the low and climax of the movie story has always been a mystery to me.

Throughout my adult life, classical music has been a constant companion. For decades, I've listened to an all-classical radio station during my commutes, and it's helped me unwind and reflect. It soothes me and instills a sense of calmness.

I couldn't afford instruments or lessons when I was young, but that didn't stop me from developing a deep musical intuition. I joined a high school chorus and learned to read music, which led me to enjoy vocals. My nickname was "The One Verse King." I'd hum catchy melodies on repeat, often driving my family nuts.

My family shares this musicality. My wife, who is also untrained, sings beautifully and learns songs effortlessly. My older son played violin with grace before pausing his journey after a personal loss. My younger son, an expressive performer, blossomed into a multi-instrumentalist and a singer, forming bands and performing at weddings. Music reaches people in countless ways. Within my own family, each of us connects with music in a unique way. We have varying musical preferences due to our different exposures, yet we all share a passion for it. This diversity is typical in most families, particularly in culturally rich environments like the United States. What I find interesting is what brings these different musical tastes to fruition.

Like many in the baby boomer generation, I grew up with radio, LPs, and cassette tapes. But I also became accustomed to CDs, iPods, and now streaming. Each technology opened new doors to the music world. I've experienced firsthand how my musical instincts adapt and endure.

I believe musicality isn't reserved for professionals. It's a birthright we all share, an invisible thread connecting us to ourselves.

Music Enthusiasts

Shared musical experiences often foster a sense of connection and community. At concerts, for example, the boundaries between performer and audience blur. There's a shared energy, a resonance, that transcends individuals.

I remember going to an Elvis concert in Austin, Texas, in the mid-1970s. Elvis gave everything he had, guitar in hand, legs shaking, hips rolling. The crowd responded as one, swaying, screaming, reaching out. The atmosphere was electric. I even saw someone faint from the sheer emotional intensity. It felt like we were all part of one body, breathing and moving together. I experienced a similar moment at a Tom Jones concert in 2022. Despite performing with a cane at age 82, he still made the audience roar. The scarves, the charisma, it was all still there. It wasn't just a concert. It was a shared emotional release. One cannot help but wonder what brings these people and their emotions together.

I've seen this kind of resonance across genres and generations, from the Glenn Miller Orchestra swinging through Northern California to the Beach Boys' fans lighting up stadiums with cell phone flashlights. Taylor Swift's 2023 Era Tour evoked that same collective pulse. It's as if the crowd becomes a single, rhythmic entity expressing its collective rhythms.

There's science behind this, too. Humans are naturally empathetic and prone to imitation. When we're immersed in music together, our brains sync up in ways that make emotional responses contagious. That's why live music can feel so powerful (Chapters 7 and 8 will explicitly address this synchronization from a neuroscientific perspective.)

That said, not all music settings promote this kind of resonance. I love classical music, but formal concert etiquette, like sitting still and remaining silent, can feel rigid. Flash mob performances, in contrast, bring classical music to life in public spaces where people can respond freely and interact with the music. The synchronization and sense of unity are when the musical magic truly happens. I even got a few goosebumps when a soprano joined one of the flash mob events.

Music also brings people together through shared identity and purpose. College fight songs, military anthems, and even advertising jingles tap into our collective musical instincts. From Top Gun's rousing score inspiring Air Force recruits to cafés curating mood music, we see how music shapes emotion and connection everywhere.

We might not always realize it, but music is a constant presence. It colors our holidays, enriches our social gatherings, and links us to something greater than ourselves. Whether we're singing along in the car or swaying with thousands at a stadium, music reminds us that we are never truly alone.

Music Lovers in Special-Need Demographics

If musicality is a basic human trait, how does it appear across our diverse population? People vary in abilities, life experiences, and preferences. Yet music touches nearly all of us, often in very personal ways.

Interestingly, music has been shown to help people facing cognitive challenges or health issues. For example, aging often causes memory decline, but music can trigger memories and emotions in those with Alzheimer's or Parkinson's disease. Studies indicate that even when verbal memory decreases, musical memory can stay surprisingly strong.

Music therapy is emerging as a powerful tool for rehabilitation. It can help stroke victims recover speech, assist individuals with autism in social interaction, and reduce anxiety in people with trauma. These effects aren't magical; they demonstrate how deeply music is wired into our brains and emotions. The link between music and health is so strong that the highly acclaimed American soprano Renée Fleming is involved with several organizations to promote understanding and use of music therapy as a healing and wellness tool, especially in brain health.

On the opposite end of the spectrum, a small percentage of people, about 1.5%, have a condition called amusia. These individuals do not enjoy music due to differences in brain connectivity, especially between sound-processing and reward-related regions. Although rare, amusia highlights the complex interaction between sensory and emotional brain systems in musical appreciation.

There are also musical savants: people with extraordinary musical ability, often emerging despite or even alongside other cognitive impairments. Their talents remind us that musicality can manifest in many forms, not all of which are typical.

Some people even experience music as color, a condition called synesthesia, where sounds evoke vivid visual imagery. It's rare, but it provides a glimpse into the powerful interaction between our senses. It also suggests that musical ability is so natural that it could spill over to other senses.

Importantly, musical access isn't always equitable. Some children grow up without access to instruments, music classes, or even the time to listen, due to life circumstances. But music often finds a way. I remember classmates from farming communities who would return to school humming puppet show melodies they'd heard the night before. Although exposure might have been limited, musical joy found its way.

In short, while our musical experiences vary due to biology, circumstance, or opportunity, our shared capacity for music remains a unifying human trait. Music is not just for the trained or talented. It's for everyone.

Human's Long Love Affair With Music

From a historical angle, music has been with us since the beginning. Today, when we hear Beethoven's Ninth Symphony, we travel across time. Musical notation and recording technology preserve emotion and intent from long ago. Yet music existed long before such tools.

Flute ensembles made from bird bones, some dating back over 9,000 years, suggest the existence of sophisticated musical scales and even ensemble performances. These weren't crude instruments. Some had seven holes, implying full scales and expressive capabilities.

Going even further back, archaeological evidence shows that humans were crafting and playing musical instruments at least 60,000 years ago. But before instruments, we had our bodies, hands for clapping, voices for singing, feet for stamping the rhythm. Biological musical instruments came first, followed by mechanical ones.

Why did we make music so early in our history? It likely served many purposes: ritual, emotion, communication, identity, and cohesion. Around fires, in ceremonies, during dances, music brought people together and helped them express what words couldn't. Some speculate that it may have played a role in our evolution by strengthening group bonds and emotional intelligence. Musicality cannot be more natural if evolution is involved? This point has been debated for ages. Hopefully, this book provides some clarification that we will contemplate toward the end of this book.

The creation of instruments reflects humanity's ingenuity. We've always sought new sounds, some to imitate nature, others to express inner feelings. Today's composers, such as Hans Zimmer, continue this legacy by inventing instruments and techniques to capture the modern emotional landscape. These inventions are believed to have originated from the composers' ability to mimic natural sounds with emotional reflections expressed in music.

In summary, music is an ancient, universal, and remarkably resilient art form. While Chapters 6, 7, and 8 will explore the brain's role in musical experience, this chapter emphasizes something just as profound: our deep-rooted, emotional, and instinctive connection to music.

Part II: Musical Obsession

The follow-up question beyond musicality is then: what happens when this inborn musicality spills over into obsession, shaping not just individual lives, but entire cultures and civilizations?

What is our true connection to music? It's much more than entertainment. For millennia, people have turned to music to express joy, sadness, love, grief, defiance, and their sense of identity. Music says what words alone often can't. It reaches deep into our emotional core and remains there as a backdrop of the human species and culture.

This section explores music not as a collection of songs but as a cultural force. It's embedded in our history, lifestyles, politics, business, and even medicine. It's both backdrop and centerpiece, an invisible presence that shapes how we live and who we are. We'll look at two overarching perspectives. This part, Part II, will focus on the emergence of a music culture and how shared musical languages and values became integral to our identity. The next part will look into music's integration into various aspects of life, including history, movies, pop culture, religion, business, and technology.

The ubiquitousness of music through space and era warrants an extended exploration to delve into the richness of the relationship between music and humans. Let's begin.

Music Culture

Talking Music

Whatever we are obsessed with, music needs to be shared with others. How do we convey what we feel about music, an experience that vanishes the moment it ends? It's not easy. We often turn to metaphors and imagery. We say music is "bright," "full-bodied," or "uplifting." These descriptions may sound poetic and fleeting, but they serve a purpose: they give form to something intangible.

Metaphors are our way of making sense of abstract sensations. Just as wine lovers describe taste using terms like "crisp" or "velvety," music lovers describe sound with equally vivid language. When someone calls a performance "captivating" or a melody "hypnotic," they're inviting us into their emotional world.

Take Don McLean's "American Pie", a lyrical journey through the shifting culture of the 1960s, we might call it bold, clever, or nostalgic. But those labels only scratch the surface. Metaphors give us a language to connect with others about how music feels, not just how it sounds.

Not everyone is so well-versed in music, but we try to delve into our intellect and emotions to describe what we hear. Talking music is an outward expression of our feelings to share; it is an internal engagement with something we like or dislike. Our minds are already de facto occupied by music when communicating with fellow music lovers.

<u>Musical Culture Scope</u>

Music culture encompasses a range of styles (genres) and elements (attributes). The breadth and the depth are as diverse and expansive as human culture. Picture it like a matrix: genres on one side (classical, jazz, country, etc.) and musical attributes on the other (melody, rhythm, harmony, and emotion, etc.) Table 2.1 is a simplified matrix that any music lover can fill in.

Genres evolve, combine, and influence each other. Aaron Copland blends classical and country. Gershwin fuses jazz and classical orchestral music. Across cultures, musical genres mix to reflect the complexity of identity and expression.

Musical Culture Matrix					
	Structure	Melody	Rhythm	Harmony	Lyric
Classical					
Country					
Big Band					
Rock and roll					
Jazz					

Table 2.1. Visualization of the complete scope of musical culture. The genres are listed from top to bottom, whereas the attributes can be the descriptors of the genre. The musical culture scope table can be very inclusive when every music lover fills in the blanks.

Attributes, on the other hand, help us understand what makes a piece of music unique. Is its rhythm steady or syncopated? Is its texture layered or simple? Is the emotion calming or exhilarating? By identifying these attributes, we improve our understanding and our ability to describe what we hear and feel. The details of the matrix, shown as Table 2.1, are not the main focus; instead, they provide a framework to highlight the diversity of musical cultures. Readers can fill in the gaps based on their own experiences and preferences.

Music's Interactive Nature

Music isn't just consumed; it's shared, performed, and responded to. It relies on interaction. Applause and standing ovations energize performers. Musicians adjust to each other's timing during a string quartet. Audiences scream, cry, or sway together in response to what they hear.

Common to all cultures, the more interaction among participants, the deeper the culture takes root and the more widespread it becomes. Regarding musical participation, the creation side requires talent, training, creativity, and vision to succeed. The consumption side, however, is accessible to everyone and feels much more natural. Although these two sides may seem distinct (this distinction is somewhat arbitrary for discussion), they constantly interact to inspire music through creation, performance, or appreciation.

The interaction can happen in various settings. It can occur at concerts when the audience applauds and gives ovations, showing approval, while the musicians feel uplifted by the positive feedback. Even composers have the opportunity to observe the audience's reactions and critiques. Performer-audience interaction often takes place during live shows. Some notable examples include performances by the Beach Boys and Tom Jones in the Bay Area. The younger audience screams loudly just at the sight of them, while older generations, who were teenagers during Tom's peak, still dance and shout at each of his signature moves. Interaction between artists and their audiences is timeless.

Musicians, artists, and performers often interact with each other during performances as well. For example, Goethe once described four players in a chamber music piece as communicating through body language, visual cues, and music. The complex interplay among the players involves adjusting to each other's pace, rhythms, timing, emphasis, and mood shifts, resulting in a give-and-take in timing and rhythm that creates an improvisation-like atmosphere of intimacy within small groups. This improvisation functions, in a way, as a democratic group working toward consensus.

Fandom is a way for artists and their audiences to connect; it values authenticity in musicians' personalities, music, and daily lives. Social networks help foster closer bonds between fans and artists. In that sense, fandom reflects the genuine emotions shown through the music and the musicians.

Music's Popularization

In the past, close musical interaction among people was common. Music and dance brought communities together during celebrations or pre-hunting rituals, which were central to tribal life. Imagine everyone gathered around the fireside, moving in rhythm to the drumbeat. The chanting and dancing built energy as the drummer set the tempo, leading to the ceremony's climactic moments. There is something instinctive and deeply rooted in our subconscious; there was no real difference between you and me, the elite and the general public, or the shamans and their followers.

This intimacy changed over time. As tribal rituals evolved into more formal musical traditions, such as church services or classical music, the connection between music creators and audiences began to weaken. For thousands of years, an artificial barrier kept elite composers apart from ordinary listeners. In early churches, for example, solemn chants were performed without allowing the congregation to join in. Similarly, ancient Chinese courts used music to regulate official ceremonies and even created a special department to oversee music's role in government and diplomacy. Ordinary people had no access to these "elegant" forms of music, which used the "elegant" scale. This Chinese ceremonial music, based on this scale (discussed in Chapter 3), reflected efforts to maintain order. Standardized scales were enforced partly because music naturally spread among the public, diversifying and challenging official control. The human obsession with music ensured it circulated from elite circles back into the public realm, a pattern seen worldwide.

For centuries, the divide between creators and consumers persisted. Classical music illustrates this clearly. In France, the royal court once enjoyed performances, and this cultural setup persists today: attend a modern symphony or opera, and you'll notice two distinct groups —musicians and composers on one side, and the audience on the other.

This rigid divide began to fade for two main reasons. First, the Enlightenment weakened traditional class hierarchies, promoting more egalitarian societies. These cultural changes influenced all aspects of life—including music. For example, Italian opera in the 18th century, influenced by Enlightenment ideals, started blending folk elements into high-class compositions.

Later efforts to bridge the gap persisted, though not always successfully. Take Aaron Copland's *Fanfare for the Common Man* as an example. Its bold trumpet fanfare evokes a sense of royal grandeur, meant as a tribute to ordinary people. Yet, ironically, those hearing it in concert halls were often aristocrats, far removed from the lives of the real "common man."

Today, modern technology has become the second major force in bridging the gap between the two music communities. It has succeeded where social reforms only partly did. Music is now accessible to almost everyone; streaming platforms, digital media, and global connectivity ensure that anyone can listen to any song at any time. Beyond active listening, we are constantly surrounded by music through various media, including radio, television, films, and the Internet. Whether we notice it or not, music permeates our environment. In a way, this universal access restores music to its original purpose: a shared human experience that crosses barriers of class, status, and geography.

Part III: Music And Its Many Intersections

Music and Politics

Music doesn't exist in a vacuum. It intersects with politics, ideology, power, and protest. Sometimes it reflects the status quo. Sometimes it challenges it. Either way, music can be a powerful tool for expressing and shaping social values.

Think about national anthems. These are auditory symbolism—music that represents identity and loyalty. In times of celebration or sorrow, people often turn to these anthems for comfort and a sense of togetherness. Songs like "The Star-Spangled Banner" or "La Marseillaise" aren't just melodies; they are highly meaningful and emotionally powerful.

However, music also serves as a form of protest. From Bob Dylan's protest songs to Kendrick Lamar's critiques of systemic injustice, artists have long used music to speak truth to power. These songs don't just entertain; they provoke thought, inspire action, and give voice to the unheard.

Even purely instrumental music can carry ideological meaning. Beethoven's Ninth Symphony has been used across political lines—from communist rallies to pro-democracy protests. The music itself is so emotionally powerful that it becomes a vessel for whatever significance people assign to it.

In some governments, music is censored because of its power to inspire dissent. That alone demonstrates its significant influence. Conversely, government-sponsored music might promote patriotism or conformity. In any case, it's clear that music is more than just art; it's a form of influence.

In everyday life, music reflects personal identity. Genres like hip-hop, punk, or folk often do more than just represent a style—they embody community, history, and values. Whether at a protest or on a Spotify playlist, the music we select reveals what we believe or at least what we want to be part of.

Ultimately, music reflects and influences society. It heightens our principles and our conflicts. Whether inspiring change or fostering a sense of community, music is political—not because it's partisan, but because it holds power.

Music has been extensively exploited throughout history. It has been deliberately used as a tool for colonialism and imperialism. It helped establish imperial dominance in subordinate lands through religious conversion. For example, the Portuguese converted about 100,000 Japanese between 1541 and 1577, when Father Organtino Gnecchi wrote from Kyoto that "if only we had more organs and other musical instruments, Japan would be converted to Christianity in less than a year." For Japan's sake, it was fortunate there were not many organs. However, Japan could not resist the trend and formed a Westernized orchestra in the early 20th century.

When Western powers controlled global materials and human resources, they told their subordinate states that Western music was considered cultured, scientific, emotional, and "universal." Music was used to enhance the strength and superiority of these countries and contributed to colonialism by dividing races and perpetuating racism. Fortunately, both colonialism and imperialism have diminished since WWII. However, music still carries racial and ethnic connotations, which are waiting to be eradicated.

Postcolonialism and imperialism allowed Western music to maintain global influence. A few years ago, while traveling to Machu Picchu, I saw street vendors trying to attract customers to their souvenir shops by playing the well-known central theme of "El Condor Pasa" on pan flutes. I was surprised and confused by its presence there until I Googled its origin (yes, I had Internet access in the middle of Machu Picchu). Peruvians consider that melody to be their second national anthem. This "native" Peruvian music was initially composed in the Western style in 1913 by Daniel Alomía Robles, inspired by traditional Inca melodies. The arrangement and adaptation by Simon & Garfunkel, which crossed national boundaries and reached the top of the Billboard chart in 1970, demonstrates the international appeal of good music and its widespread reach.

The familiar tune "doe, a deer, a female deer..." popularized by the American movie "The Sound of Music" about an Austrian musical family, was sung by nearly all preschoolers worldwide in the 60s. The American folk song "Dreaming of Home and Mother" by John Pond Ordway became one of the most memorable Chinese folk songs at the beginning of the 20th century. Who knew a Chinese folk song could originate from North America?

Music and Ideology: Not Just a Historic Footnote

Political science and philosophy have long overlooked the role of music in shaping ideology; for a long time, many writers argued this, until people like Jean-Jacques Rousseau took a different approach. Rousseau was the rebel of his era, circa 1740-1760, opposing everything associated with the prevailing establishment. According to Rousseau, music and politics are two sides of the same coin, and music has historically served the rulers of political and ideological hegemony. However, it can also serve to neutralize, resist, or question power, which was a significant goal of the Enlightenment. Therefore, music should not be seen simply as a footnote or an accessory to political thought and action, but rather as a core element. The key idea is that music is deeply connected, through its aesthetics, to ethical judgments and social order. From a less philosophical perspective, it has played a significant role in historical, social, and political contexts in more practical ways.

During my college years, I found football games to be among the most exciting activities. Most college teams have marching bands and fight songs. I recall the University of Texas marching band continuously playing The Texas Fight Song, or The Eyes of Texas, during the game, supposedly demonstrating the University's pride and tradition. I was amazed at how the audience often formed a cheering wave from one end of the stadium to the other, following the rhythm and prompts of the music. The band's performance inspired the audience and, more importantly, the players, regardless of the competition's outcome. It is fantastic if the home team wins, but there is also a sense of camaraderie even in defeat. College football games are not the only infectious occasions. In the seventies, Bum Phillips' Houston Oilers had the team song "Houston Oilers #1." It is a bit juvenile for professional sports, but it united the Oilers' fans in sports bars and living rooms throughout the Houston metropolitan area. The US Marine Song recounts the battles at Montezuma and Tripoli. Along with the marching music, the cadets cannot help but embody the spirit of the Marines.

Most national anthems embody broader ideologies and can have significant political effects. For example, 'Nkosi Sikelel iAfrika' is recognized as modern South Africa's national anthem. Composed by Methodist school teacher Enoch Sontonga in 1897, it was initially sung as a church hymn (though the religious aspect is a separate topic) before evolving into a symbol of political defiance against the apartheid regime. It united the people against apartheid, ultimately contributing to its downfall. 'Nkosi Sikelel iAfrika' played a vital role in fostering what Archbishop Desmond Tutu called the rainbow nation. It has transformed the idea of egalitarianism from mere potential into reality.

Reflecting on history, music has long supported politically and ideologically dominant powers, as Rousseau acknowledged. One example dates back at least 2,500 years in China, where music was considered a vital part of effective governance. The protocol department, one of six departments at the time, set standards and conducted official ceremonies in state affairs. Officially called "elegant music," it directed rituals offering tribute to the divine for favorable climates and plentiful harvests. By around 600 BCE, the development of musical scales and instruments indicated that the universal "perfect fifth" had been in use for quite some time. Consequently, the concept of diatonic scales was well established, forming the foundation of "elegant" music (the subject of Chapter 3). Ceremonial music, along with its accompanying lyrics, was primarily designed along ideological lines, aimed at wielding power for governmental purposes or inter-state events.

Music and its Globalization

There is no denying that Western music and styles dominate modern music worldwide. How did this happen? Colonialism was the central ideology adopted by Western culture, primarily through corporations like the Dutch West India Company and the British East India Company (EIC), following the start of global trade in the early 17th century. This global commerce began when Western Europeans discovered new trade routes at the end of the 15th century. Since then, imperialism has taken further steps to control colonized and conquered territories. Both practices contributed to cultural dominance worldwide, and the widespread presence of Western music reflects this cultural supremacy.

Initially, music was used either consciously or unconsciously for nostalgic reasons. For example, the earliest record of European music in the New World states, "The *Pinta* leads the procession, and her crew is singing the Te Deum (a religious chant). The crews of the *Santa Maria* and the *Nina* join in the solemn chant, and many of the rough sailors wipe tears from their eyes." Christopher Columbus recorded these words in his journal on October 12, 1492, as his three ships landed in America. This marked the start of Western music's dominance over the colonies when it first arrived in America. As colonization grew, Western music mainly aimed to recreate the home environment for the colonists amid the widespread dominance of colonialism, imperialism, and Christian conversion.

Since 1694, the port city of Calcutta had the British East India Company exporting materials such as silk, opium, and indigo, while importing Western goods, including opera. In 1775, Calcutta proudly opened its first European theater, the Calcutta Theatre. By the early 1800s, a large number of British colonists, mainly from the upper classes, had settled in Calcutta. They were eager to maintain their European traditions and established European theaters, connecting opera to Imperial Victoria Calcutta. Most musicians and singers were imported from Europe, with Calcutta often serving as a stop on their tour. A few local Indians began to imitate and eventually became important contributors to Western music, leading to a more localized opera scene in Calcutta. Music such as Sati ki Kalankini fused

Bengali theatrical traditions with Western musical influences, emerging in the late 19th century.

In the late 19th century, Shanghai's high society and Western expatriates founded the Shanghai Municipal Orchestra, which served a nostalgic purpose similar to that of the Calcutta theater, while also drawing a steady Chinese audience. The musicians and audience of the orchestra shifted from a purely Western identity to a blend of Chinese and Western influences. This shift reached its peak in 1990 when the orchestra performed a special invited concert at Carnegie Hall in New York to mark its 100th anniversary. A century earlier, Tchaikovsky had conducted the first concert at Carnegie Hall. A hundred years later, Chinese musicians took the stage for the first time, earning praise from the 2,300 audience members present. Local music critics praised the Shanghai Symphony Orchestra the next day as "a world-class orchestra." This concert represented the complete transformation of music from Western to hybrid, marking the end of the old colonialism and imperialism that began 300 years ago.

Music and Film/TV

Music and movies/TV have always shared a symbiotic relationship. Movie music can be seen as a unique subculture that dates back to the late 19th century, when motion pictures first appeared. The earliest documented use of music in cinema happened on Dec. 28, 1895, when the Lumière brothers, considered the first filmmakers, tested the commercial potential of their initial films. That screening took place in Paris and was accompanied by a live piano performance. The first film with a recorded soundtrack was shown on November 17, 1908, featuring Camille Saint-Saëns' music for the film "The Assassination of the Duke of Guise." Around the same time, the French company Le Film d'Art promoted adding music to every film. Then came the 1926 movie Don Juan, directed by Alan Crosland, which premiered at the Warner Theater in New York. It was the first film to feature a synchronized music score and sound effects. The soundtrack for Disney's 1937 animated film Snow White and the Seven Dwarfs was the first to be commercially released as a film soundtrack.

The rest, as they say, is history. Today, we have specialized music composed and played specifically for movies and TV to enhance their stories. Movies and TV shows without music are like artwork without souls. It's hard not to notice how Superman's score excites the audience with its opening theme by John Williams. Who can forget the music from The Lion King by Hans Zimmer? Zimmer is so inventive that almost anything can become part of his new instruments, and his music and sound effects blend seamlessly with films.

Movies need music to evoke the emotions connected to visual effects. At the same time, music requires visual elements to amplify the feelings embedded in the tune and lyrics. TV shows also have close ties with music. The theme song of "M.A.S.H." creates a somber mood symbolic of the cruelty of war. The "Cheers" theme song makes you want to visit a bar where everyone knows your name. The "Big Bang Theory" conveys a nerdy vibe that is often looked down upon by many. Its theme song, however, recounts the early history of the universe, being both musical and even educational.

Movies also reshape the landscape and culture of music, especially classical music in both composition and structure—much like the past—without strictly following composers like Hayden, Mozart, and Beethoven. They include classical music in pop movie culture in surprising ways. Who would have thought that the Boston Pops would combine moviemaking with classical orchestral performances? The Danish National Symphony Orchestra's rendition of the theme song from "The Good, the Bad, and the Ugly," decorated with movie-themed decor, might seem to stray from traditional classical music. However, due to Clint Eastwood's spaghetti Westerns' enduring pop culture influence, this classical version is becoming part of pop culture.

The flip side of this symbiotic relationship is how music evokes emotions in movies rather than being subordinate to the film. I recall the Superman theme gradually building up Superman's power and ultimately reaching a crescendo as he absorbs energy from the sunlight. I believe everyone experiences their own emotions while listening to it. But that is also the beauty of music; like other pure art forms, it invites individual interpretation of the artists' intent in creating the artwork.

Even animated films have included classical music, as shown by Mickey Mouse conducting Johann Sebastian Bach's Fugue in D minor. It brings to life the magic imagined by Mickey Mouse and, naturally, the audience. Without Mickey's story, classical music might not have gained as much popularity among the general public. Or think of "Ride of the Valkyries," which is used as the opening for the movie "Apocalypse Now."

<u>Music and Pop Culture</u>

Not only does music have an intimate connection with movies, but it also permeates other aspects of life and pop culture. During a cruise trip to Alaska, there was a tribute to the British Invasion of the US (focused on the music, of course). One song that caught my attention was the Beatles' "Lucy in the Sky with Diamonds." I can't help but think about the iconic status of "Lucy" in anthropology. In this context, Lucy refers to the small-bodied woman who lived about 2.5 million years ago. Her skeleton was discovered in Africa by anthropologist Donald Johanson in the mid-1970s. Lucy was named after the woman linked to the song, which played repeatedly in the tent throughout the day after the discovery.

In high school chemistry, while studying how chemical elements undergo oxidation and reduction, we were asked to memorize the oxidation-reduction potentials of elements in order from highest to lowest. The teacher taught us a song starting with "Lithium, Rubidium, Potassium, Cesium, Barium, ..." Music helps with chemistry, at least for me. The fact that lithium is at the top of the list explains why it is used in modern batteries everywhere.

<u>Music and Religion</u>

Music was once seen as a spiritual monument, with its creation believed to reveal divinity through artists' minds. Beethoven and Mozart are said to have shared similar beliefs. Music hints at the existence of an all-powerful being who communicates from above through artists' closer connection to the deities. It has also been effective in gaining loyalty. Although the mystery surrounding music has lessened, its role in religious ceremonies and solemn occasions remains a significant part of today's church services.

When and how religious practices began are probably unknowable; however, there are a few indications that worship of the divine started very early. The earliest evidence of physical monuments or temples dates back to an underground cave with ring structures, believed to have been built by Neanderthals around 176,000 years ago in southern France. This discovery was made in 1990. The cave site has DNA remnants indicating that Neanderthals inhabited it for more than 100,000 years before modern humans (Homo *sapiens*) even set foot in Europe. Additionally, there is evidence of fire use from deposits found on the surfaces of stalagmites covering the ring structure. While traveling around Şanlıurfa in southeastern Turkey, known for its pistachio production irrigated by the Euphrates River, I visited an ancient temple called Göbeklitepe, built 12,000 years ago. The megalithic stelae feature carvings of god-like creatures, animals, and human figures that form multiple rings. Some ceremonies were probably performed around these rings. The Stonehenge monument, located near London, was built roughly 5,000 years ago. It had astronomical significance, as the principal axes aligned with the time of the winter solstice, signifying that the worst part of the year is over.

The purposes behind these structures remain shrouded in lost history, yet they could have served as monuments around which ceremonies were performed. Could music have been part of the ceremonies and rituals paying tribute to their dignities or deities? It is reasonable to assume a large group of people would gather around these sites to voice their thoughts of the day, either in a public forum or during ceremonies surrounding the physical monument. One wonders what the group did by the fireside beyond talking, dancing, and planning for the next day's activities. Anthropologists believe fire and music must have been present during their construction and use.

We were not present to witness what transpired, but we can see remnants of these rituals in the Indian rain dance performed by native peoples of the Americas, accompanied by chanting and drumbeats. I visited an Aboriginal tribe site in the '80s; one of the tribe members danced and played music on an instrument made of a string pulled between his teeth and his hand. With his other hand, he plucked the string while varying the pulling strength, producing different tones for others to sing along. The dance celebrated the gods, thanking them for granting a prosperous harvest year, all while circling a roaring campfire that emerged from a large fire pit. As I later learned, the Aborigines are Austronesian people who practice long-held traditional rituals, paying homage to their god, who grants them good fortune. They are considered the earliest modern human migrants from Africa, arriving as early as sixty thousand years ago, reaching the general South Pacific islands and as far as Australia. I have not encountered similar musical instruments since. Human ingenuity knows no bounds when it comes to creating musical instruments in tribute to their deities.

There is limited documentation of music culture before Greek times in Western culture. Nonetheless, music was present as early as 500 BCE. The accumulation of human experimentation with instruments led to the Greek era, when records of music began to appear in the form of artifacts, such as decorative motifs on their containers. This period also marks the Greeks' recognition of the power of music to enhance the expressive meaning of language or words through their dramas and plays.

Greek music was adopted by the Romans around 500 CE, during the decline of the Roman Empire and the rise of the theocratic era. The church came to oversee every aspect of human life, including music, and since then, music has become almost inseparable from religion. The dedication to the church produced mesmerizing hymnal music like Gibbons's "Almighty and Everlasting God," which dominated the European music scene for centuries.

The Renaissance marked a revival of Greek culture—literature, art, and music—following the Dark Ages and medieval times, which is why it's called the "renaissance." Although Western culture gradually moved toward secularism after the Enlightenment, today's Western music still retains a significant religious element because of the enduring connection formed over centuries.

Soul music is a prominent example that traces its roots to religion, especially Christianity. This popular genre of music originated within the African American community across the United States in the late 1950s and early 1960s. It has roots in African American gospel music and was influenced by the then-popular rhythm and blues (R&B) trend. Soul music became known for both dancing and participation, with U.S. record labels like Motown, Atlantic, and Stax playing key roles during the Civil Rights Movement. Notable artists and groups include Aretha Franklin, Maxine Nightingale, The Manhattan Transfer, The Stylistics, and The Emotions. Soul music, along with a major part of U.S. pop music, remains closely linked to religion.

The influence of religion on music, or vice versa, has shaped the forms and styles of today's music. However, music should not be mistaken for possessing deep, divine meanings. As will be explained later, music's effects on human emotion and psyche relative to religion are not mysterious and can be understood through biology and science.

<u>Music and Lifestyle</u>

Music culture is a lifestyle that constantly evolves, much like fashion. Loving and living the Louis Vuitton lifestyle signifies a distinct aesthetic; driving luxury cars reflects a particular lifestyle; appreciating coffee is also seen as a lifestyle; and wine tasting is often regarded as a lifestyle. Anything that sparks widespread interest and trends that capture public attention, like music, is considered a lifestyle. All lifestyles tend to change over time. It was once unimaginable in classical music that anyone could envision soul music, modern pop, or hip-hop. We can see the future of music (we can speculate, though) just as we can predict next year's fashion trends. Future music will inevitably go beyond today's categories and genres.

This new music lifestyle trend transcends gender, race, genre, and more, evolving into a diverse array of lifestyles. The key link between music and lifestyle is how it is experienced. Public consumption—through radio, headphones, and boomboxes—brought musical experiences into the open. The barrier between music and daily life, heightened by the ritual of the concert hall, dissolved as music increasingly became part of people's personal daily environment. Yet, these changes were simply preludes to the transformation driven by digital technology; therefore, the story of changing music continues to develop alongside lifestyle, and vice versa.

From generation to generation, one of the most common complaints among older generations about new music is that the younger generation's music is declining. These persistent cycles have existed for hundreds of years, at least, in Western culture. I was on a cruise ship from San Francisco to Hawaii in late 2022, and after spending many days at sea, the cruise director had to invent activities to keep the travelers entertained. One of the games involved playing music trivia from various periods. Our impromptu team of travelers, composed of a few retirees, correctly identified 39 out of 40 songs and artists from the 50s and 60s pop music. The same group was able to identify only one out of 40 for the music genres of the 80s. Just 30 years have passed since music subcultures in the US began to diverge. Music has acted like a fashion statement for generations; in other words, it reflects the lifestyles and cultures of its time.

Music and Technology

Music and instruments have always been closely linked to technological advances. Originally, humans likely used their vocal cords to produce different tones for expressing emotions and conveying meanings, especially when messages needed to go beyond simple knowledge and be communicated through language. Early proto-music probably served as a way to reinforce the memory of emotional messages.

The exact time of humanity's first tonal musical instrument may be unknowable. The first person who either intentionally or accidentally discovered that blowing through a straw, hollow bones, or rolled-up leaves could create pleasant pitches must have also learned that plucking the strings on a stretched bow can produce tones similar to those of humans. They found that these crude instruments could better reproduce the tones, giving proto-music much-needed repeatability. Like myself, they must also have been very curious that blowing into these pipes in different ways could produce distinct tones, similar to the differences in our voices between males and females. To fill the tonal gaps dictated by the length of the tubes, holes are drilled along their length, thus creating scales between octaves.

The speculation may have centered on how our ancestors created the first verifiable instrument: a flute made from a cave bear femur, discovered in Cerkno, northwestern Slovenia, dating back between 40,000 and 60,000 years. The recreation of that flute, based on 3D tomography, can play modern music with reasonable accuracy; however, it only plays minor scales, which are likely defined differently than they are today. With the bone well-polished and the hole precisely positioned, it's clear that making the flute required a vast amount of knowledge gained through generations of trial and error, much like our high-tech inventions.

It is reasonable to assume that many instruments would emerge once it is recognized that various natural materials can be shaped into instruments. For example, stones are the main components of percussion instruments. The discovery of ancient stone chime ensembles shows their excellent tone and consistent scale. In ancient Chinese culture, they even standardized the materials used for music. These are called the eight tones and are made from metal, stone, clay, leather, string, wood, gourd, and bamboo. With the development of these musical technologies, humans could combine vocal and instrumental sounds, shaping the music landscape as we know it today.

The next major invention was musical notation. Its purpose was to record music in a way that it could be read, replayed, and, hopefully, preserve the spirit of the original composition. This invention is important because it keeps music alive forever and passes it down through generations. In theory, by playing from musical notation or sheet music, one should be able to recreate the original intentions of the musicians, including which instrument to use, which holes to cover on the flute, how long to hold notes, and the intended tempo. However, we still cannot exactly reproduce the music composed 200 years ago by composers like Beethoven or Bach. For example, one cannot replicate the performance of Nocturne in E-flat Major, Op. 9, No. 2 by Chopin from the sheet music, as there is only one Chopin.

Nevertheless, the notation marked a significant technical advance that supported music and its development over the past few millennia. But something was still missing until we could capture the fleeting music experience from the air into a recorded medium that we could revisit and enjoy again. This technological breakthrough began as early as the 1860s. That invention worried Arthur Sullivan, a well-known English composer, who once said, "I can only say that I am astonished and somewhat terrified by the result of this recording (referring to the event): astonished at the wonderful power it has developed, and terrified at the thought that so much hideous and bad music may be recorded forever." However, he could never have imagined the power of recording in terms of its authenticity and its capacity to spread music without limits.

This writing does not aim to provide a detailed history of various technological innovations for recording, but rather to highlight related technological advances: analog and digital technologies/recordings. Analog technologies aim to record and reproduce every detail of the music, including all its nuances, hence the term "analog." The production and distribution of music in analog form played a key role in popularizing music from the 1880s through the 1960s until digitized music became commercially viable. Since then, digital technologies have taken over the distribution of music that now extends across the Internet cloud.

Analog technologies, despite the dominance of digital technologies, remain a crucial part of music because our ears cannot directly perceive digitized sounds. Modern analog technologies focus on improving acoustic and sound engineering, as well as the auditory experience. We can also enhance music using technology to elevate it beyond the original from the listeners' perspective. For example, audio devices like speakers or headsets can deliver sound directly to the inner ear without interference (with noise cancellation) or obstruction from the environment. Moreover, these technologies can be customized to fit the listener's personal preferences or specific needs, all made possible by analog technologies.

Digital technologies can convert music to any desired quality, distribute it worldwide, convert it back to analog upon reception, and optimize it for human listening. Depending on the digitization methods used, a single piece of music can be encoded in millions of bits of data. The music can be reproduced with 100% accuracy, with no errors in the bits. Digital storage is ideal for preserving the sound of music forever without any deterioration.

What possibilities exist for music to collaborate with technology in the future? One of the many benefits is that they can help push the boundaries of creativity in music. Technology can replicate every musical instrument available, such as keyboards. Even a novice like me can play music with the help of preset rhythms and chords, allowing me to focus on the main themes. Technology can also perform instrumental music like a professional if given the right input. When provided with complete music sheets, it can perform an entire symphony like an orchestra.

Additionally, technology enables machines to learn vast amounts of musical knowledge from the Internet. With that, an intelligent machine can now compose alongside human composers if instructed correctly. As technology advances toward integrating human emotions and feelings, it could become creative and rival master composers. The extent of artificial intelligence's influence on music will be a topic of future discussion. Technology has helped initiate the music culture, co-evolving over thousands of years and ultimately pushing boundaries to create new music for the future.

Music and Business

The relationship between music and business has undergone significant changes. In ancient times, music was primarily used for rituals and community gatherings, rather than for commercial purposes. Still, back then, musicians traded their art for support and favors as business transactions. By the medieval period, minstrels and troubadours performed for royal courts in exchange for protection and backing—an early form of today's music industry.

The Renaissance and Baroque periods marked another milestone: music publishing. Composers made deals with publishers to distribute sheet music, while the rise of opera in 17th-century Italy led to the creation of commercial theaters, turning music into a primary industry.

The industrial era accelerated this trend. Recording technologies built a mass market for music in the 19th and 20th centuries, followed by radio, record labels, and the modern music industry. Today, music is monetized through live concerts, streaming platforms, licensing, and merchandise.

How big is the music business today? The global music market is estimated to be worth $30.7 billion in 2024 and is expected to reach $45.65 billion by 2029, growing at an annual rate of 8.54%. For perspective, the worldwide semiconductor (IC) industry, which powers smartphones, cars, computers, TVs, networking equipment, and medical devices, is valued at about $422 billion. The fact that the music industry can be mentioned alongside the vital IC industry is both astonishing and fascinating. Although music is smaller in absolute size, its ability to compete with such a significant sector is impressive for what is, fundamentally, a musical art form.

Consider Taylor Swift, whose blend of artistry and business savvy has made her a global superstar and a billionaire. Named *Time*'s Person of the Year in 2023, Swift's concerts create a substantial economic impact. Two shows in Santa Clara, California, brought in $30 million for the local economy through ticket sales, restaurants, hospitality, and retail. Her U.S. tour of 150 shows generated billions in national revenue, and her international expansion further boosted that impact. In Singapore, her concerts contributed nearly $1 billion to the country's GDP of $600 billion.

Swift's influence goes beyond music. Her appearance at National Football League games increased merchandise sales by 50%, drawing in new fans who hadn't watched the sport before. She even secured multi-million-dollar endorsement deals with Singapore Airlines, merging cultural influence with economic strength. Swift's success highlights a key truth: great music fuels great business, because music is more than just entertainment; it's a driving force of culture and commerce.

Music and Health

The healing power of music has been acknowledged since ancient times. Ancient Greek doctors used flutes and lyres to soothe patients, aid digestion, and even promote sleep. I was reminded of this during a visit to Asklepion, a well-known healing sanctuary in Pergamon (modern-day Bergama, Turkey), dating back to the 4th century BCE and named after Asclepius, the Greek god of healing. The site integrated medicine, spirituality, and recreation. Patients bathed in mineral waters, had their dreams interpreted, and watched theatrical performances in a 3,500-seat amphitheater while music was used to ease their suffering.

This ancient wisdom now has scientific support. Since the 1980s, research has shown music's positive effect on cognitive health, especially for people with Alzheimer's, Parkinson's, and aphasia. Music can revive memories long believed lost, assisting Alzheimer's patients in recalling personal experiences and even dance steps from years past.

Why does this work? Music activates alternative neural pathways when primary routes are blocked by conditions like Alzheimer's, where amyloid plaques disrupt normal signaling. The brain's redundancy, its ability to rewire and compensate, supports this recovery. Music acts like physical therapy for the mind: rhythmic patterns and familiar melodies stimulate dormant circuits, promoting neural plasticity, the brain's remarkable capacity to adapt.

Even unfamiliar music can ignite interest, improve mood, and encourage motor activity, aiding in the formation of new connections. Essentially, music acts as a neurological workout. Just as repetitive movement strengthens injured muscles, engaging with music enhances alternative brain pathways by promoting dendrite growth, thereby facilitating the formation of new connections.

As neuroscience advances, tools like fMRI reveal the dynamic relationship between music and brain plasticity. These discoveries confirm what ancient cultures believed: music has healing properties. Far from mystical, its power is rooted in biology, a topic explored more thoroughly in Chapter 7. Music's role in health, once based on anecdote, now rests on a strong scientific foundation, offering hope to patients around the world.

<u>Music and SETI</u>

Music has even reached space. The Voyager Golden Record, part of the S.E.T.I. project initiated and supported by figures like Carl Sagan, was launched into space in 1977, carrying music from different cultures as a greeting to potential extraterrestrial life.

SUMMARY

We've come full circle. This chapter started with a question: why is humanity so obsessed with music? The answer is in a feedback loop —where *musicality* fuels *obsession*, and *obsession* nurtures *musicality*.

Musicality is part of our biological makeup. It resides in our brains, bodies, and cultures. From babies bouncing to rhythms to elders moved by melodies they've known for years, we're all wired to connect with music. It's not a luxury or an afterthought—it's a fundamental aspect of being human.

That innate capacity fuels our obsession. We create, consume, and live through music. It expresses who we are, reflects our values, and unites our communities. From prehistoric bone flutes to pop songs on TikTok, music evolves with us and because of us.

And the obsession, in turn, sharpens our musicality. We seek new sounds, invent new instruments, and create new traditions. We teach and learn music in schools, stream it nonstop, and let it mark the moments that matter most—birthdays, weddings, victories, and goodbyes.

Music isn't just entertainment. It's experience. It's memory. It's emotion. It's a connection. It's history. And most importantly, it's universal. Is there something underlying this universality? In the upcoming chapters, we'll explore how this universal obsession sheds more light on our musicality from biological, physical, and neuroscientific perspectives.

BIBLIOGRAPHY

1. "Gimme Paleolithic Shelter," Zack Zorich, *Archeology Magazine*, 2016.

2. "Music Market Size & Share Analysis - Growth Trends & Forecasts (2024 - 2029)," *Mordor Intelligence,*

3. https://www.mordorintelligence.com/industry-reports/music-market-landscape

4. Wikipedia: https://en.wikipedia.org/wiki/History_of_sound_recording

5. "Music and Mind: Harnessing the Arts for Health and Wellness," Renée Fleming, *Viking,* 2024.

6. "Breaking the Silence: Music's Role in Political Thought and Action," J. Street, *Critical Review of International Social and Political Philosophy,* **10**, 321-337 (2007) https://doi.org/10.1080/13698230701400296

Chapter 3

Scales' Universal Nature

Most multilingual individuals have opportunities to experience music from diverse cultures. They likely find it easy to listen to music from one culture, switch seamlessly to another, and enjoy it just the same. I know I can. These individuals often discover that modern music across cultures utilizes diatonic or chromatic scales for performing, singing, and composing.

These scales are based on standardized audio frequencies and were created using long-standing traditional methodologies. In 1939, an international conference in London recommended setting the standard A4 at 440 Hz (cycles per second), now commonly known as concert pitch. The International Organization for Standardization (ISO) officially recognized A4 = 440 Hz as "Concert Pitch" in the 1950s and further formalized it in ISO:1975. The typical pitch pipe used for tuning vocal and string instruments is set to this A4 pitch. This standardization allows instruments to play together, even when some are difficult or impossible to tune to their original pitch due to material and construction limitations. Today, musicians compose, read, and perform using scales based on this standard. The global music scales could not be more uniform, and this commonality among cultures is unmistakably clear. One might wonder how we arrived at this point. Is there something in our nature that made this commonality inevitable? Or is it a result of the ongoing cultural globalization that has been happening long before recorded history?

The musical scales were not very uniform during the period leading up to the start of modern globalization, which began in earnest in the early 16th century, shortly after Columbus discovered the "New World." Before this era, the "Old World," generally including Eurasia, India, and Africa, was culturally divided into Eastern and Western regions, with the Western usually referring to Europe and West Asia, and the Eastern to China's sphere of influence. Although Eastern and Western cultures are fundamentally different, ethnomusicologists and music theorists have found that, despite some differences, these musical cultures—particularly in scales—share a common scientific and mathematical foundation, making them essentially similar, and even identical. As it turns out, these intriguing similarities stem from physical and biological reasons across vastly different cultures, as would become evident after the elaboration of Chapters 6, 7, and 8.

The systematic and scientific efforts to address these issues began early in recorded history. Notable historical figures include Pythagoras from Greece, Guan Zhong from China, and Rikpratisakhya from India, all active around 4-7 centuries BCE. They observed music's hidden regularities and orderliness, establishing rules based on their observations. From these empirical theories, they developed the pentatonic, diatonic, and 12-tone chromatic scales, which have lasted for thousands of years. These theories and practices have been independently developed, leading to similar results, sophistication, and internal consistency.

We are now better equipped to understand the physics of acoustics and the scientific tools needed for measurements and verification. We recognize several similarities between Eastern and Western music and realize that the scale's many perceived complexities are largely artificial. Once the cobwebs are cleared away, it becomes easier to reconcile theories from different cultures. After all, beneath the commonalities lies how the scales arise from our innate musicality, which is rooted in common biological responses to music.

Octave and Scales

When comparing musical scales across cultures, two key similarities stand out. First, all musical instruments produce tones with multiple harmonics related by octaves, regardless of the materials or designs used. The second commonality relates to how octaves are divided into scales. Although variations exist across cultures, the notes on the scales are connected by simple ratios of string lengths in string instruments within the octaves.

Any naturally sustained sounds from vocal cords or instruments include a dominant fundamental tone at a specific audio frequency along with their harmonics. By naturally generated tones, we refer to sustained single-syllable sounds that keep their pitch either vocally or through instruments. These sounds may be a string instrument played at constant tension, flutes played at consistent lengths, or cymbals being struck. These harmonics occur at frequencies that are multiples of the fundamental frequency, such as double, triple, or higher. Figure 3.1 is a spectral graph of my attempt to sing the note of C3. As can be seen, my voice contains the dominant tone at 263 Hz (approximately, of course) and at least two more components at twice and thrice this fundamental frequency. They are the harmonics of the C3 sound I made.

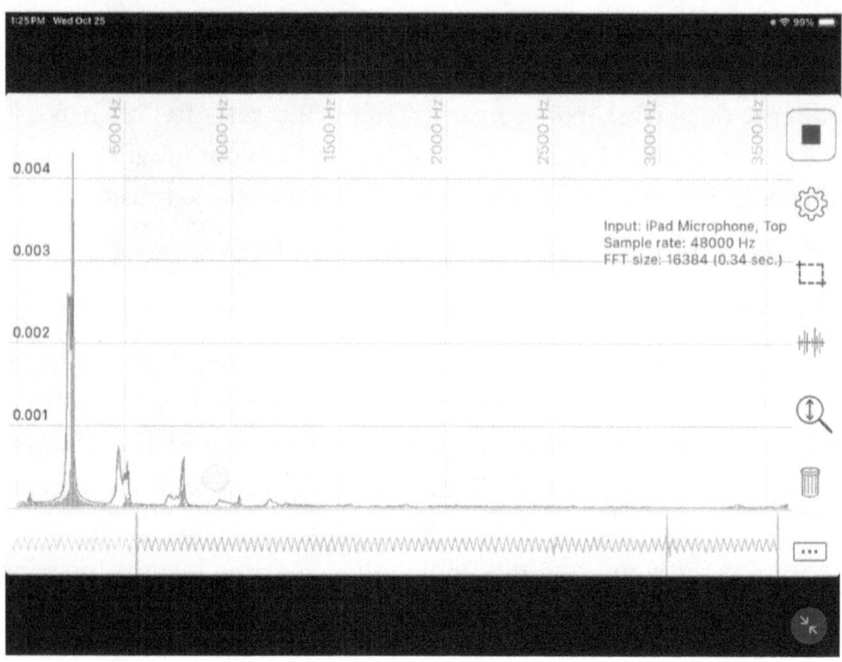

Figure 3.1. The spectral graph of a human voice singing a C3 tone. Notice the fundamental tone at the right frequency as well as the harmonics at twice and thrice the fundamental frequency.

Each harmonic relates to the fundamental tone by octaves. Since they are naturally produced, humans have become accustomed to them throughout our evolutionary history, like other primates, making them generally pleasing or at least not objectionable to our perception. This means the same instrument can sometimes produce sounds at double or triple the fundamental tone by chance if the conditions for generating higher harmonics are met. In physics, these specific conditions are called initial or boundary conditions. For example, when you blow into a section of a flute without holes, it likely produces the flute's fundamental tone. However, if you blow into the flute at a different angle, it might produce a shrill sound that could be one of the harmonic tones. Those familiar with guitars may be familiar with playing harmonics. When any string is slightly touched at the midpoint (near the twelfth fret), it produces the second harmonic as the primary vibration mode, making the dominant tone. This second harmonic becomes dominant because the initial condition favors it, while the light touch suppresses the fundamental tone. Even when you blow a whistle harder, you might unintentionally hear harmonic tones one or more octaves higher in pitch.

These higher frequencies' harmonics are where higher octaves are built upon. This phenomenon likely occurred naturally but not consistently in the early days. When it happens more often, we have tried and succeeded in replicating the conditions under which harmonics are produced regularly. Experimentation may show that these higher harmonics are consistently present across various musical instruments. The starting pitches, or audio frequencies, of these instruments can be located anywhere on the audio spectrum that humans can hear, mainly determined by the materials and dimensions of the instruments. Nonetheless, the presence of octaves remains constant.

After becoming more familiar with the octaves, we might have found it monotonous that each instrument could only produce its fundamentals and harmonics for human enjoyment. How can we introduce some tones between the octaves? When tightly pulled, hunting bowstrings emit a distinct twang. Observers also noticed that a higher pitch would emerge if part of the long string were pressed to mimic shorter strings. This tone can exist between the octaves as long as the string isn't shortened by more than half. Thus, the length of the string can determine the tone relative to the fundamental notes. It became possible to combine several strings to cover various octaves and the tones in between, leading to the creation of the lyres, harps, pianos, guitars, violins, and many more.

We have also seen pan flutes, where flutes of different lengths are tied together to produce various tones. This is richer in sound than a single tube that can only produce one tone and its harmonics.

But why use many flutes when we can simulate different flute lengths by punching a few holes in one flute? They were undoubtedly the predecessors of clarinets, trumpets, bassoons, horns, and so on.

Providing a comprehensible list of global music scales across cultures and history would be a huge task and wouldn't effectively clarify the topic. Countless musical traditions exist worldwide, each with its own unique scales and playing modes. Wikipedia can be a helpful reference for an extensive list of possible scales; however, covering every scale imaginable is not possible. This book will focus only on well-known scales, including Western major and minor scales, Indian ragas, Arabic maqamat, and Chinese twelve-tone scales.

Before exploring the topic further, we must recognize that all scales are based on octaves that can start at any audible frequency. A4 at 440 Hz is a standard for convenience and compatibility only; this will be repeated when appropriate.

The recognition that naturally produced sounds have multiple octaves is instinctive to humans and other primates, as it is neurologically supported. According to a study by Bendor and Wang in 2005, "pitch sensitive neurons" in a non-human primate brain are a specific group of neurons located in an area of the auditory cortex that respond selectively to the fundamental frequency of a sound. This allows the primate to perceive pitch even when the fundamental is "missing" in a sound, such as a missing fundamental harmonic complex, demonstrating pitch constancy in animals. By extension, humans likely share this trait, and the ability to recognize octaves is fundamental, forming the basis for how they are divided, rooted in our evolutionary instincts. More details will be provided in Chapter 6 when neural science identifies the location of the pitch-sensitive function in our auditory cortex. Therefore, recognizing repeating octaves is a natural phenomenon that everyone experiences. Consequently, it should not be surprising that it is a key aspect of our musicality.

Western Music Scales

Diatonic Scale

Western music has a long history of ongoing invention, innovation, refinement, solidification, standardization, unification, flourishing, and maturation into its current form. One of its most important innovations is the scale. In modern Western music, the diatonic scale is the most common.

The word "diatonic" comes from the Greek word "diatonikos," which means "through tones." A diatonic scale has seven notes: five whole steps and two half steps arranged in the pattern whole-whole-half-whole-whole-whole-half, then returning to the root note an octave higher. This pattern defines the Ionian mode, also known as the major scale.

Consider the white keys on a piano spanning from C4 to C5, the seven notes—C, D, E, F, G, A, and B—constitute the diatonic scale. The eighth note, C5, is an octave above the root note. This eight-note range is the origin of the term "octave," which comes from the Latin *octavus*, meaning eighth. Today, we use this term to describe any situation involving a doubling of appearances or frequencies for different purposes. For example, electrical circuits with resistors, capacitors, and inductors often have frequency-dependent characteristics described in terms of variations per octave.

In the 11th century, an Italian monk and music theorist named Guido of Arezzo revolutionized music notation. He created the solmization system using syllables: *do, re, mi, fa, sol, la,* and *ti.* These were derived from a Latin hymn, *Ut queant laxis,* sung in the Gregorian chant. Today, Guido's system is widely used to teach sight-singing and is made even more famous thanks to the musical *The Sound of Music.*

You might wonder why pianos have black keys and why there is space between the whole notes and some half notes on the white keys. What exactly are whole notes and half notes? As mentioned earlier, scales can start at any tone or pitch, and any set of seven notes can serve as the root key for creating a complete diatonic scale. However, changing the root key from C to D, for example, means the new scale can no longer follow the familiar whole-whole-half-whole-whole-whole-half pattern. This is where the difference between major and minor scales comes into play. The complexity would disappear if we started with a chromatic scale, which also serves as the foundation for diatonic scales.

One wonders how diatonic scales are created and integrated into the audio space across octaves. The clues for achieving this lie in string and pipe instruments from early times. The pitches are known to change based on the lengths of the strings or the pipes. Therefore, it makes sense to vary the lengths of the strings to produce scales. Combining strings of specific lengths forms the foundation of instruments such as the piano, guitar, lyre, and harp. Even the mouth-pulled instrument I mentioned earlier (Chapter 2) operates on the same principle, using string tension to change the pitch.

The same idea applies to wind instruments, such as those commonly seen in flutes. The full scale between octaves is covered with holes drilled at strategic positions. Our fingers cover these holes to imitate pipes of different lengths, thus different tones. The same principle is the foundation of any wind instrument.

<u>Chromatic Scale</u>

When an arbitrary string length defines a root key, the subsequent scale is achieved by varying the lengths of the instrument strings. We are most familiar with the perfect fifth, which occurs when the string length is reduced by one-third while the new pitch remains within the same octave. Since the audio frequency of a string is inversely proportional to its length, this one-third reduction in length results in a 50% increase in audio frequency connecting the C and the G notes as a pair of perfect fifths. The exact process is repeated six times to complete a full octave, completing the diatonic scale. Continuing this process five more times, a total of 11 times, lands the 12th tone almost on another octave. The twelve perfect fifths operation creates the chromatic scale.

The diatonic and chromatic scales, along with their naming conventions, can be visualized using a music theory tool that originated in the late 1600s and early 1700s: the circle of fifths. This tool assists composers and songwriters in theorizing and modulating music from the Baroque era. The circle of fifths shows the relationships among the 12 major and minor keys, how they overlap, and how they share notes. It also displays the related minor keys

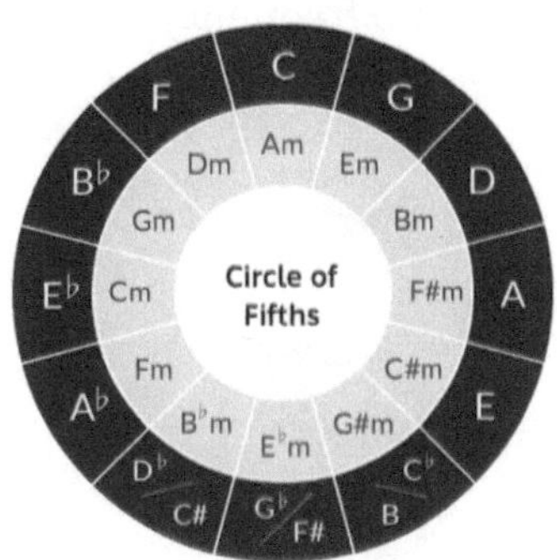

Figure 3.2. The circle of fifths, a visual representation of this relationship, organizes pitches in a sequence of perfect fifths, highlighting closely related keys.

and the six diatonic chords in each key. This process can be visualized as shown in Figure 3.2. The circle begins at the top with note C, with the following notes moving clockwise: G, which is a perfect fifth above C. Continuing this sequence by adding the next perfect fifth clockwise produces D, A, E, B, F#, C#, Ab/G#, Eb/D#, Bb/A#, F, and back to C. Semitones space the twelve notes generated this way, and each is either sharp or flat relative to its neighboring notes. These twelve pitches and their audio frequencies, created through perfect fifths, are distributed relatively evenly in a multiplicative manner. However, they do not bring the twelfth note exactly back to the next octave. We will address this discrepancy in the next section.

One alternative way to visualize the twelve notes is by arranging the 12 semitones in ascending acoustic frequencies in a full circle, as shown in Figure 3.3. When each semitone is connected to its perfect fifth counterparts across the circle with a line, the twelve lines form a dodecagram, a 12-pointed convex star. The cyclic nature of this diagram allows connections of perfect fifths to start from any note (or audio frequency). As well, starting the dodecagram from a C note is simply a convention.

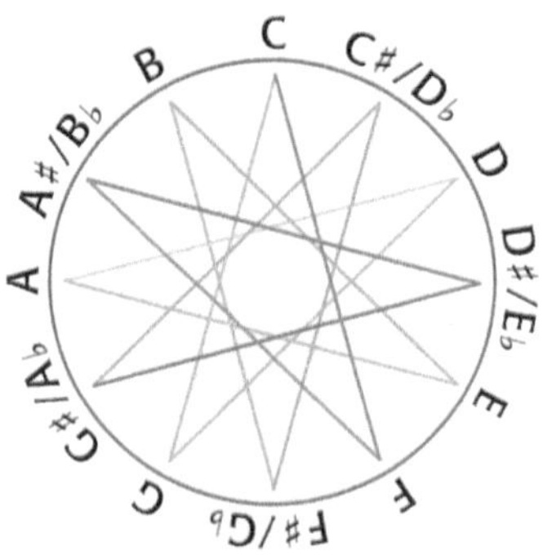

Figure 3.3. This "musical dodecagram" is a visual representation of the twelve notes of the chromatic scale arranged in a 12-pointed star or dodecagram.

The circle of fifths and the dodecagram help demystify music theory for composers and learners alike. I often use them myself, especially the dodecagram, when I need a refresher on keys, sharps, flats, and scales.

Pythagorean Comma

Octaves are precise and predictable. Doubling the frequency of a tone always produces the next octave. However, this does not apply to the process of successive perfect fifths, as it does not mathematically bring you back to the exact octaves. After twelve iterations of the perfect fifth, you will approach a nearest octave, but not on top the octave. The difference between this and the actual octave after seven iterations is the Pythagorean comma, named after the mathematician credited with its discovery.

Mathematically, this imperfection is expressed by the ratio of $(2/3)^{12}$ to $(1/2)^{7}$ of 1.013643265 instead of 1.00 for perfect overlap, leading to a deviation of 23.46 cents in musical terms, where a cent is $1/100^{th}$ of a semitone. Although small, the difference is perceptible to most ears.

This imperfection meant that early scale systems had to make compromises. Some intervals sounded slightly off, especially in keys farther from the root. Over time, musicians and theorists developed alternative scale patterns to distribute the discrepancy more evenly. These eventually led to the equal temperament system, where the octave is divided into 12 exactly equal semitones, removing the comma's cumulative error.

Interestingly, this 23.46-cent discrepancy also appears in the tuning systems of other cultures, like ancient China, showcasing how different traditions independently faced the same musical challenge. As we will see later, nearly identical mathematical methods have been used to address this issue across cultures, especially between Chinese and Western scales. This convergence suggests that Western and Chinese cultures share similar skills in both music and mathematics.

Chinese Twelve-Tone Scales

Westerners historically believed that the pentatonic scale was the primary scale used in Chinese music. This misconception may stem from the fact that the well-known Chinese folk song "Jasmine Flower" features a pentatonic structure. John Barrow introduced "Jasmine Flower" to England in 1792 during the George Macartney mission to China, which occurred during imperialistic times when non-Western cultures were often considered primitive and simple.

One-Third Loss Gain Algorithm (OTLG)

In fact, China developed a methodology to create the 12-tone scale as early as around 1,000 BCE. The pentatonic scale was only part of a more extensive set of the full scale used on limited occasions. Ancient Chinese musicians and theorists devised a well-structured approach to generate the twelve-tone musical scales, which are identical to the Western chromatic scale.

The method of dividing the audio spectrum among octaves in ancient China dates back at least 2,700 years. Guan Zhong, a figure known for legalism and authoritarian rule who lived around 700 BCE, mandated that the One-Third Loss Gain (OTLG) algorithm be used to create the scales for official music. The idea that string instruments can produce different tones within an octave and that the frequency of a tone decreases as the string length increases (assuming all other string properties are equal) probably existed long before Guan Zhong's era. This understanding likely gave him the confidence to formalize the OTLG process as an official regulation.

The simplest way to understand this is to visualize how the OTLG functions and creates the 12-tone scales based on string lengths. It starts with a musical instrument with a string length of 81 units, which is the tone of Gong (the start of the Chinese scale solmization, or the root note "do" in Western music). The first step involves reducing the string length by one-third, resulting in a new length of 54 units. Playing this shortened string produces the tone of Zhi (or "so" in Western music). As we know, this is recognized as the perfect fifth in Western music. The OTLG then increases the string length by one-third, bringing it to 72 units, producing the tone called Shang (or "re" in Western music). Repeating this process of decreasing and increasing the string length allows us to complete the diatonic scale with the five notes of Gong, Shang, Jue, Zhi, and Yu, corresponding to do, re, mi, so, and la. These five notes make up the familiar pentatonic scales. The note names correspond to the same solmization system as the pentatonic scale in Chinese music.

Around 240 BCE, government officials advanced the OTLG operation beyond the fourth step. The fifth and sixth steps stayed within the same octave, constituting the seven notes of the diatonic scale. The seventh step of OTLG is supposed to cause a one-third length loss; however, it causes the tone to spill into the next octave. But if you lengthen (gain) the string twice from that point, effectively bringing the tone back into the original octave, you can continue the OTLG process from there. After the 11th iteration of OTLG, we have nearly reached the next octave. The entire set of 12 tones forms the 12-tone temperament, similar to the process of executing the perfect fifth and obtaining the 12-tone chromatic scales in Western music. Figure 3.4 visually demonstrates this OTLG process and its corresponding Western music scale, using A3 on the piano as a reference for calculations. The figure also details the steps involved in the OTLG process.

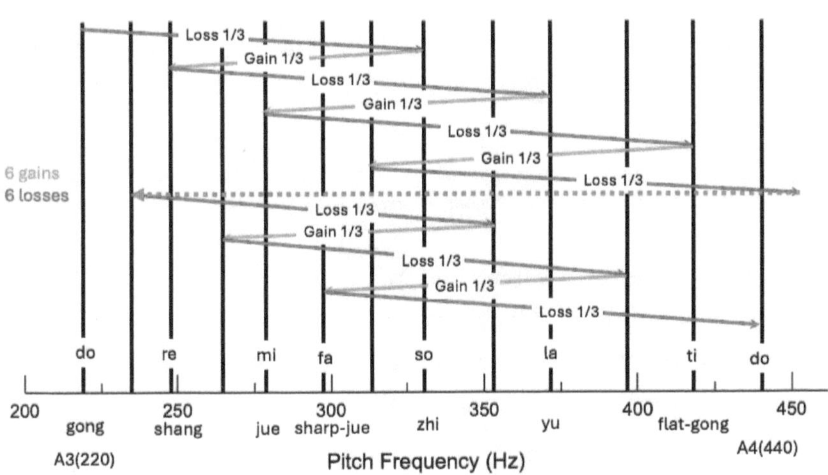

Figure 3.4. *A graphical representation of the Chinese scales constructed from the OTGL algorithm. Inherent in this graph are all variations of Western scales: the pentatonic scale, the diatonic scale, and the 12-note dodecatonic scale (also known as the chromatic scale).*

Table 3.1 summarizes the previous descriptions and compares the corresponding modes from Chinese and Western musical scales.

Tone Names	Equal Temperament frequencies	Frequencies from OTLG	String Length units	Chinese Tone Names	OTLG Operation	Italian syllables, Ionian Mode	Chinese Pentatonic Scale	Elegant Mode, Same as Lydian Mode	Crisp Mode, Same as Ionian Mode	Feasting Mode, Same as Mixolydian mode
A3	220.000	220.000	81.000	HZ1	HZ1	do	Gong	Gong	Gong	Gong
A3#/B3b	233.082	234.932	75.852	DL	RB - 1/3 then + 2					
B3	246.942	247.500	72.000	TZ	LZ + 1/3	re	Shang	Shang	Shang	Shang
C4	261.626	264.298	67.424	JZ	YZ + 1/3					
C4#/D4b	277.183	278.438	64.000	GX	NL + 1/3	mi	Jue	Jue	Jue	Jue
D4	293.665	297.335	59.932	ZL	WS + 1/3	fa			sharp Jue	sharp Jue
D4#/E4b	311.127	313.242	56.889	RB	YZH + 1/3			flat Zhi		
E4	329.628	330.000	54.000	LZ	HZ1 - 1/3	so	Zhi	Zhi	Zhi	Zhi
F4	349.228	352.397	50.568	YZ	DL - 1/3					
F4#/G4b	369.994	371.250	48.000	NL	TZ - 1/3	la	Yu	Yu	Yu	Yu
G4	391.995	396.447	44.949	WS	JZ - 1/3					sharp Yu
G4#/A4b	415.305	417.656	42.667	YZH	GX - 1/3	ti		flat Gong	flat Gong	
A4	440.000	446.003	39.955	HZ2	ZL - 1/3	do	Gong	Gong	Gong	Gong

Table 3.1. This table summarizes the processes of the one-third-loss-gain (OTGL) method, the solmization of the 12 tones in Chinese scales, and the equivalent modes of scales for Chinese and Western scales for comparison.

OTLG demonstrates a direct way to identify the 12 notes in the chromatic scale based on their frequencies, as shown in Figure 3.3. The concept of audio frequency was not well understood in China until the 18th century, except in the context of string lengths. However, a graphical aid similar to the Western circle of fifths has been used since the Tang Dynasty, around 750 CE. This is shown in Figure 3.5, where the five smaller circles represent the pentatonic scale with the notes do, re, mi, so, la, or gong, shang, jue, zhi, and yu, corresponding to the tones of C, G, D, A, and E as shown in Figure 3.1.

Figure 3.5. This Chinese circle of fifths has been used since the Tang Dynasty, around 750 CE, as a learning aid. The five circles represent the pentatonic scale, with the five notes corresponding to the tones of C, G, D, A, and E as shown in Figure 3.1.

Each of the twelve chromatic tones produced by the OTLG was assigned an official name as early as 2,500 years ago: they are HZ, DL, TZ, JZ, GX, ZL, RB, LZ, YZ, NL, WS, and YZH, representing their solmization. The root key is always called HZ, which can start anywhere on the sound spectrum. The other eleven tones follow the OTLG rules and adjust accordingly. HZ literally means "Bell" or, more specifically, "Bronze Bell," which probably looks like Figure 3.6 and officially sets the standard tone for the root key, similar to how Bach aimed to standardize various scales in his "Well-tempered Clavier" book. It also shows where the official HZ frequency should be, just as ISO 16 defines concert pitch as A4 at 440Hz.

Figure 3.6. A photograph of the standard Bell, more specifically, "Bronze Bell," which formally establishes the standard for the root key for Chinese music as early as the fifth century BCE.

Chinese musical scales are used for string and wind instruments and serve as the foundation for making percussion instruments like bells and stone chimes. During an archaeological dig in Hubei Province, China, a 5th-century BCE tomb was uncovered that included a complete set of bells and chimes. The main bell in the group is likely the HZ bell, as shown in Figure 3.7. The ensemble consists of 65 bells spanning six octaves.

As shown in Figure 3.8, a complete set of stone chimes, tuned to the full chromatic scale, was also found in the same tomb, aligning with the bell scales covering three octaves.

Based on the 12-tone scale, the Chinese made slight adjustments to the diatonic scale to boost its variation and richness. They created three diatonic scales for official occasions: elegant, crisp, and feasting. Elegant is used for the most solemn and formal events; crisp is designated for light-hearted and joyful celebrations, while feasting is reserved for state dinners and grand banquets.

In 11th-century Europe, churches served as centers for music, where chants and hymns, such as Gregorian Chant, were performed for religious purposes. Musical talents, mainly located in churches or monasteries at that time, developed seven slightly different diatonic scales based on variations in the spacing between tones. Table 3.2 lists these seven modes along with the corresponding scales that connect the Church modes with Chinese scale modes constructed from the 12 semitones. The Ionian mode represents the most common major scale, following the sequence of whole and half steps: w-w-h-w-w-w-h or the 2-2-1-2-2-2-1 semitone pattern. The varying spacings and modes shown in Table 3.2 provide the basis for major and minor scales.

Figure 3.7. *This photograph shows the recreation of an archaeological find from a 5th-century BCE tomb. The complete set of bells comprises 65 bells covering six octaves. The central bell in the ensemble is the HZ bell, functioning as the root key, as shown in Figure 3.5.*

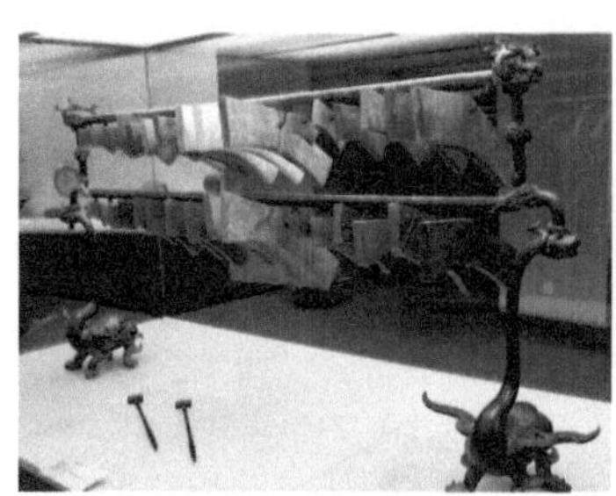

Figure 3.8. *A complete set of stone chimes, tuned to the full chromatic scale, was also discovered in the same tomb, consistent with the bell scales covering three octaves.*

Church Mode	Number of semitones in between							Chinese Style
Ionian	2	2	1	2	2	2	1	Crisp Scale
Dorian	2	1	2	2	2	1	2	
Phrygian	1	2	2	2	1	2	2	
Lydian	2	2	2	1	2	2	1	Elegant Scale
Mixolydian	2	2	1	2	2	1	2	Feasting Scale
Aeolian	2	1	2	2	2	1	2	
Locrian	1	2	2	1	2	2	2	

Table 3.2. *The seven church modes of the Western diatonic scale and the three modes of the Chinese diatonic scale.*

As mentioned earlier, after performing the OTLG 12 times, the tone "ALMOST" lands right on the next octave. It has been known since ancient times that "ALMOST" is not the same as reaching the top of the next octave. This flaw is officially called "HZ Not Brought Back to Origin (or Octave)," or HNBBO for short. Table 3.1 uses A3 at 220 Hz as the HZ for simplicity. It shows that the OTLG process results in the next supposed HZ at a higher octave being 446.003 Hz instead of 440 Hz, missing by 23.46 cents. This misalignment flaw is common in both Western and Chinese musical scales. Using A3 as the starting point is only for illustration; using any other notes or audio frequencies to initiate this OTLG process will produce the same flaw, exactly 23.46 cents.

Other Scales

While the scales from other cultures may differ, their fundamental principles and practical origins often remain consistent. Here is a summary of several different scales.

Indian Music Scales

Indian classical music uses a detailed system that combines parent scales and their related families. Like Chinese and Western scales, each pitch has a specific name. The seven pitches in Indian music are called svars: Ṣaḍja (Sa), Ṛiṣabha (Re), Gāndhāra (Ga), Madhyama (Ma), Panchama (Pa), Dhaivat (Dha), and Niṣāda (Ni).

In Hindustani (North Indian) classical music, the most common way to classify a raga is by using ten parent scales called thaats. A thaat consists of a seven-note scale, including each of the seven notes. Similar to Western music, the Indian music scale also has 12 semitones. This scale can be further divided into 22 shrutis by adding extra tones between the semitones.

Arabic Music Scales

Maqam scales in traditional Arabic music are microtonal and do not follow the twelve-tone equal-tempered tuning system used in modern Western music. However, most maqam scales feature perfect fifths, fourths (or both), and octaves.

Javanese Music Scales

Javanese Gamelan music has been extensively studied since the 1600s, possibly due to the West India Company's influence in that region. Metallic cymbals and other percussion instruments with well-tuned scales characterize the music. It resembles the Chinese official standard stone chimes and bells, which were tuned to the chromatic scales around the 5th century BCE, mentioned earlier in the discussion of Chinese music scales.

Across all these traditions, a common theme emerges: musical scales embody both mathematical logic and human experience. Whether through strings, bells, chants, or improvisation, cultures around the world develop systems that organize pitches in a mathematically similar way.

The Equal Temperament

Western music used the circle of fifths to create the 12-tone chromatic scale, while Chinese music employed the OTLG. Western scales encountered the Pythagorean comma problem, and Chinese scales faced the HNBBO problem. Unaware of each other, both have the same imperfection of 23.46 cents from the next octave.

From the last row in Table 3.1, starting at A3, the imperfection after the OTLG process yields 446.003 Hz (from the last row in Table 3.1) instead of the ideal octave of 440 Hz for A4. A cent is a unit of measurement for the frequency ratio between adjacent semitones. An equally tempered semitone is defined as 100 cents, leading to

1,200 cents for a full octave. Thus, each cent corresponds to = 1.0005777895... ratio in audio frequencies. Consequently, the ratio of 446.003/440 of 1.01364326 is this ratio raised to the 23.46[th] power, resulting in a 23.46 cents error for HNBBO. According to the definition, the previously mentioned Pythagorean Comma of $(3/2)^{12}/2^7 = 1.01364326$ derived from the circle of fifths method is also exactly 23.46 cents. Both Western and Chinese music have encountered the same issue using the same algorithm for over 2,500 years. What a surprise!

Is this minor imperfection significant to our hearing? The perceived impact depends on several factors. It may relate to how our ears respond to audio frequencies, considering that each hair cell (discussed in the next chapter) can detect the movement caused by air disturbance—a purely physiological issue, specifically our ability to resolve frequency differences, or resolution for short. Another factor is how we train our hearing to discern subtle differences in tones. Some of us can distinguish tones that differ by as little as 10 cents, and a difference of 23.46 cents would undoubtedly annoy most musicians.

Exploring a circle of fifths and OTLG reveals similar flaws, but the 12-note chromatic scale created this way is an excellent starting point, with "nearly even" spacing between the 12 semitones. If we were to start from what we know today, a better way to divide the octave would achieve two objectives: (1) retain the traditional 12-semitone scale as closely as possible, or at least very similar, and (2) ensure equal spacing between semitones while placing the 12th tone exactly on the next octave.

It might seem trivial that the constant multiplier between neighboring semitones, when applied 12 times, results in 2 and should be the 12th root of 2. However, this was a mathematical challenge in ancient times. The earliest recorded attempt dates back to around 400 CE, when mathematical knowledge was not advanced enough to solve this with high precision. There was a lull in efforts to address this problem until the late 16th century, when an unexpected agreement between East and West emerged on this constant multiplier. Table 3.3 highlights the most significant attempts to establish the "perfect" 12-tone chromatic scales, known as equal temperaments.

Year (CE)	Responsible Person	Profession	From	Ratio Between Successive Semitones	% to Exact 12-tone Temperament
400	Chengtien He	Government Official and Mathematician	China	1.060070671	101.0
1580	Vincentzo Galilei	Musician	Italy	18:17	99.0
1581	Zaiyu Zhu	Prince of Ming Court	China	1.059463094	100.0
1585	Simon Stevin	Mathematician, Physicist and Military Engineer	Netherlands	1.059546514	100.1
1630	Pere Marin Mersenne	Polymath	France	1.059322203	99.8
1630	John Faulhaber	Mathematician	Germany	1.099490385	100.0

Table 3.3. This table lists the most significant attempts to establish the "perfect" 12-tone chromatic scales, known as equal temperaments, spanning from the fifth century CE to the 17th CE.

Chengtian He (369-447 CE) understood that whenever an extra tone was added, he needed to increase his calculated string lengths by 1/12 of their value while performing OTLG to lessen the HBBNO impact. Although this adjustment is reasonable and sufficient, it is neither mathematically precise nor elegant and does not fulfill the two stated goals.

Vincenzo Galilei (1520-1597) introduced the multiplier of 18/17, which came from musicians' need to find the best way for the lengths of the lute to reproduce the 12 semitones. To my knowledge, he was the first to accept dissonant tones because resolving them with consonance gives our psychology a pleasant sensation (see Chapters 5, 6, and 7). However, his idea of consonance focused only on harmonics and not chords. Incidentally, Galileo Galilei, a physicist, astronomer, and the inventor of the telescope, is Vincenzo's son.

Zaiyu Zhu (1536-1611) was a prince who was not favored by the Chinese Emperor and was not assigned any major tasks or duties. He had the freedom to pursue any artistic endeavors he considered important. In addition to being a musician, he was also a dancer and an astronomer. He was dissatisfied with the one-third loss-gain (OTLG) methodology due to the aforementioned HNBBO defect. He believed that a single multiplier between semitones would be the correct approach if he could determine one. He used an abacus, a Chinese calculating device invented around 600 BCE. Using an 81-column abacus, he calculated the 12th root of 2 to 24 decimal places (1.059463094359295164561825) and published this value in his book "The Treatise on Equal Temperament" in 1581. His invention of the equal temperament chromatic scale is recognized in some academic circles as one of China's five most significant inventions (the others are paper making, printing, gunpowder, and the compass).

Simon Stevin (1548-1620) was a highly skilled scientist familiar with many different fields. He made important contributions to various areas of science and engineering, both theoretical and practical. He probably understood that the 12 tones needed to be connected multiplicatively and that a simple multiplier might be the best way to get the right spacing in an octave spectrum. He started with a value of 1.05946514, which is generally close enough, and he never aimed to improve it further.

Pere Marin Mersenne (1588-1648) was the first to clearly demonstrate that the ratios between octaves are exactly a factor of two when using strings, although this fact was already widely recognized. He also discovered that a string's oscillation frequency is (1) inversely proportional to its length (which was well known at the time), (2) proportional to the stretching force, and (3) inversely proportional to the square root of the string's mass per unit length.

The frequencies of string instruments must be carefully adjusted to align with these principles, which were already acknowledged by that era. Nonetheless, he approximated the multiplier to be = 1.05973267, which is closer to the twelfth root of 2 than Galileo's 18/17.

Johann Faulhaber (1580-1635) was more known for his work in mathematics than for music. In the mid-17th century, he also calculated the precise multiplier for the 12 semitones in an octave.

The mathematical formalism for an equal temperament scale is straightforward: the octave doubles, divided into 12 semitones exponentially by a constant multiplier. Everything feels correct now.

As shown in Table 3.3, the Western and Chinese musical worlds arrived at the same single multiplier almost simultaneously, along with the means to calculate the twelfth root of two. Was this simply a coincidence? Was there any communication between them during that time? Who deserves credit for the origin of this precise equal temperament and its methodology?

George Gheverghese Joseph from the University of Manchester cited his ongoing research showing that the Chinese mathematician Zaiyu Zhu discovered the mathematical method involving the twelfth root of 2 to create an equal-tempered 12-tone musical scale by 1581 at the latest. Simon Stevin, who is often credited with this discovery, hinted at his methodology no earlier than 1585. Marin Mersenne discussed equal temperament in his 1637 book, *Harmonie Universelle*, possibly after reviewing Stevin's work. Joseph and a colleague are currently searching for evidence that the necessary knowledge may have been transmitted through the Jesuits in a chain from Zaiyu Zhu to Jesuit scholar (1552–1610), who worked in China, to Jesuit mathematician Christopher Clavius (1552–1610), then to Niklaas Trigault (1577–1628), and finally to Stevin and Mersenne.

Johan S. Bach thoroughly understood the significance of equal temperament and promoted standardizing 12 equal divisions in his compositions. He wrote 48 preludes and fugues to showcase this in his "Well-Tempered Clavier." Why 48? Because equal temperament divides the octave into 12 semitones, each capable of starting both a major and a minor scale, creating 24 modes. He composed two pieces for each mode—one prelude and one fugue—totaling 48 works. These compositions were originally written for harpsichord or clavier but also work well on organs. Today, studying his "Well-Tempered Clavier" is almost essential for any music student.

Summary

Archaeological discoveries have uncovered a variety of ancient musical instruments, some dating back thousands of years. The Neanderthal flute (Figure 3.9), found in 1995 in Slovenia's Divje Babe cave, is considered the oldest known musical instrument in the world, with an age of at least 50,000 years.

The figure shows the reconstruction of the musical instrument. The reconstructed parts are in white plaster. An arrow marks the position of the beveled cutting edge of the mouthpiece.

The flute is made from the left thighbone of a young cave bear and has four pierced holes, probably making it the oldest flute capable of playing scales. Although we cannot definitively date this invention, the ancient origins of musical scales are undeniable. Given the age of this flute, it is very likely that our ancestors, the Neanderthals, made it, as modern humans had not yet arrived in Europe at that time, which is why it is called the Neanderthal flute. A modern reproduction, created using a three-dimensional tomography model, has been shown to produce melodic music, although it tends to sound more in minor scales. This instrument, made from the body part of a cave bear, reminds me of Jean M. Auel's book and its adapted movie "The Clan of the Cave Bears," and how vividly they may have depicted the lives of humans in that long-gone era.

From that point on, humans—whether Homo *sapiens* (modern humans) or Neanderthals—created countless instruments from readily available materials for various purposes. Although they may differ significantly, octaves remained a consistent feature in each one. Our ability to hear allows us to easily identify octaves, which are inherently abstract and absolute, reflecting our evolved physiological and intellectual capabilities from ancient times. The ability to use our hearing and intelligence to introduce tones between octaves is even more fascinating, considering that we independently developed musical scales that are strikingly similar across vastly different cultures.

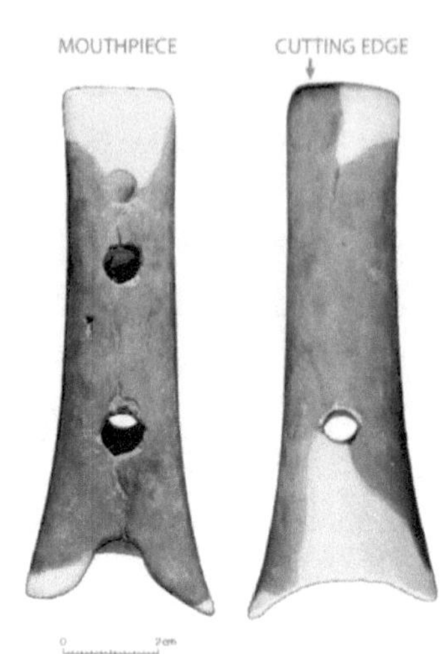

Figure 3.9. A flute discovered from Divje Babe cave in Slovenia is recognized as the oldest known musical instrument in the world, dating back at least 50,000 years. It is crafted from the left thighbone of a young cave bear and features four pierced holes, likely making it the oldest flute capable of playing scales.

This chapter simplifies and clarifies these fascinating cultural similarities by relating them to basic math and natural phenomena. Earlier methods from both Western and Chinese traditions, coincidentally destined to meet, arrive at the same general idea of musical scales through perfect fifths or one-third loss-gain techniques. The complex scales and relationships between cultures can be reduced to the number of steps between octaves. The twelve-step scale is preferred because the perfect fifth and OTGL methods divide the octave evenly, with the knowledge that they are imperfect. The ideal, unblemished twelve-step scale results from a constant multiplier that evenly spaces frequencies across octaves in an exponential pattern. These scales ultimately prove to be identical across both Western and Eastern cultures.

Now that we are satisfied with the 12 equal temperaments, we could wipe the slate clean and start anew. What if we redesigned pianos or clavichords to have only 12 keys with repeated octaves, and all music was composed and played accordingly? That would eliminate much confusion and simplify the lives of music students. A few sections in this chapter might become redundant if that were the case. However, history did not unfold as we wished, and we must confront tradition and engage with approaches from different cultures that arrived at the same conclusions. Only in hindsight can we say that the math for scales is simple and should have been established this way from the beginning, rather than grappling with traditional complexity.

Finally, this chapter concentrates on the objective mathematical clarification of the commonalities that ultimately converge. The next logical step in the process examines how we hear and perceive music, as sound produced by our voices and instruments transforms into our physiological signals. Chapter 5 discusses the methods of converting acoustics into neural impulses, marking the beginning of a neuroscientific understanding of music. However, we want to lay the scientific groundwork before proceeding further. The next chapter highlights the importance of the second law of thermodynamics in our general statistical learning and predictive coding neural processes that lead to musical reward and pleasure. It also introduces an analogy juxtaposing neural responses to music with an engineering model for simplicity and clarity. We will refer to this control system model when necessary later to illustrate the concept of musical phased-naturalism introduced in the first chapter of this book.

Bibliography

1. "The Crest of the Peacock: Non-European Roots of Mathematics," George Gheverghese Joseph, *Princeton University Press*, Oct 24, 2010.

2. "The Prehistory of Music: Human Evolution, Archaeology, and the Origins of Musicality," Iain Morley, *Oxford University Press*, 2013.

3. https://www.google.com/search?q=playing+nean-derthal+flute&oq=playing+neanderthal+flute&gs_l-crp=EgZjaHJvbWUy BggAEEUYOTIICAEQABg WGB4yDQgCEAAYhgMYgA QYigUyDQgDEAAYhg-MYgAQYigUyDQ gEEAAYhgMYgAQYigUyDQg FEAAYhgMYgAQYigUyCggGEAAYgAQY ogQyCggHEAAYg-AQYogQyCggIEAAYgAQYogTSAQkxMTI1NWowajmoAgCw-AgE&sourceid=chrome&ie=UTF-8#fpstate=ive&vld=cid:8dc30476,vid:6gueZBiRK0Y,st:0

4. "The neuronal representation of pitch in primate auditory cortex," Daniel Bendor, Xiaoqin Wang, *Nature*, **436**, 1161-5 (2005). doi: 10.1038/nature03867.

5. "The Neanderthal Musical Instrument from Divje Babe I Cave (Slovenia): A Critical Review of the Discussion," M. Turk, I. Turk, M. Otte, *Appl. Sci.* **10**(4), 1226, 2020; https://doi.org/10.3390/app10041226

Chapter 4

Sciences And Engineering In Music

Statistical Learning and Predictive Coding

As briefly introduced in Chapter 1, the theoretical model of statistical learning and predictive coding (SLPC) applies to all our senses, including sight, sound, touch, taste, and smell, as well as their corresponding neural responses. SLPC is recognized as the cornerstone of modern neuroscience. It provides a framework for understanding how the brain quickly interprets the world by anticipating what's likely to happen next based on past experiences. This enables rapid adaptation to the environment, making it a central component of many current models of perception and cognition.

Statistical learning and predictive coding are closely connected. Both describe how the brain processes information, learns from experience, and adapts to its environment. These concepts have emerged from over a century of neuroscience research and advanced rapidly in recent decades, fueled by the growth of computational models and neuroimaging technologies.

Statistical Learning, SL

Statistical learning describes the brain's ability to identify patterns and regularities in the environment based on the statistical properties of sensory inputs. It is essentially the process by which we extract information from the distribution of sensory data, such as co-occurrences and associations, probabilities, and structure, from our surroundings. This type of learning is implicit and often unconscious, meaning we do not always need to be aware of the processes of learning. For example, infants use statistical learning to understand phonetic regularities of language or to recognize visual patterns of objects that frequently appear together or move in predictable ways. This process allows the brain to build models of the world based on how often and in what ways events are observed. These models are stored in our brains in simplified forms because the statistical occurrence reduces the memory capacity needed.

Predictive Coding and Error Messages, PC and PE

Predictive coding is the theoretical framework suggesting that the brain constantly makes predictions about incoming sensory input and updates these predictions with new sensory information. The brain's goal is to minimize the difference, or "prediction error," between what it predicts and what it actually receives. This process is essential for perception, learning, and action. The term "error" here does not mean mistake literally; it is a measure of the difference between the expected and actual messages.

In predictive coding, the brain is viewed as a "prediction machine" that creates a generative model of the world, forecasting likely outcomes and refining these models with sensory inputs. It functions hierarchically, with higher-level brain regions making broad predictions and lower-level regions processing sensory data, which they then relay to higher levels to update these predictions.

Interaction Between SL and PC

Top-Down and Bottom-Up Processing

Predictive coding typically describes the flow of information from higher-level cognitive processes (top-down predictions) to lower-level sensory areas (bottom-up input). Statistical learning can influence both directions of this flow. For example, patterns identified from sensory data (bottom-up) affect higher-level predictions about the world. Conversely, top-down predictions can shape how the brain interprets new sensory input.

Contextual Learning

Statistical learning refines context based on prediction errors and their processing. Your brain may have learned from past experiences that certain sensory inputs often occur together; for example, a scene and a piece of music might be remembered in a romantic context. When this context is disrupted or changed, the predictive coding system adjusts its model based on this statistical learning to better predict future events.

Contextual learning establishes the stages for feedforward and, to a lesser extent, feedback, expanding the definition and role of the SLPC. It is a high-level process that involves complex calculations and takes time to execute and activate. Therefore, they are unlikely to be reflexive responses. A simple example of this "slower" process is how we can anticipate the crescendo of a musical piece and activate our reward seconds before it occurs.

In summary, statistical learning gives the brain a basis for environmental regularities and statistical patterns. At the same time, predictive coding uses these learned patterns to forecast incoming sensory information. Together, they enable the brain to improve its predictions, minimize surprises, and update its models of the world through ongoing learning and referencing past experiences. Statistical learning shapes the priors that guide predictive coding, and predictive coding enhances these priors through continuous sensory input and error correction.

<u>SLPC Across The Senses</u>

The SLPC mechanisms are present in all sensory systems to help us anticipate future events based on past experiences, thereby improving our perception and behavior during learning and cognition.

For visual sense, the visual system uses statistical learning to recognize patterns in shapes, objects, and environments. Over time, we learn to associate specific visual cues—like shapes, colors, or motion—with particular objects or situations (e.g., recognizing a tree by its shape or color).

For tactile (touch) sensing, we learn the statistical regularities of texture, temperature, and shape of objects through statistical learning. When we touch something new, the brain uses prior tactile experiences to identify and interpret our sensations. Predictive coding allows us to anticipate an object's texture, weight, and shape based on past experiences while reaching for it, guiding our motor actions and expectations. For example, when touching an apple, the brain predicts its surface texture and roundness before fully grasping it.

Our somatosensory system (Body Awareness and Movement) depends on recognizing patterns in body movements and postures. Over time, we learn to move efficiently and anticipate how our bodies will respond to different physical actions. The brain continuously predicts the sensory feedback it expects during movement, such as feeling when walking or reaching for something. It compares these predictions with incoming sensory information to improve movement and correct errors in real time. In a recent video, I saw a young elephant calf playing with its long trunk, trying to figure out what it is used for, a perfect example of this type of SLPC.

The Olfactory (Smell) and Gustatory (Taste) senses use statistical learning to recognize odors and flavors linked to certain foods or experiences. Over time, we connect specific smells and tastes with familiar objects or situations. In the context of smell or taste, the brain predicts the flavor or scent of food before we fully perceive it, based on past experiences. For example, the smell of coffee might cause you to anticipate, even salivate over its taste, before you take a sip.

The brain uses statistical learning in the auditory system to identify patterns in sounds, such as language, music, or environmental noises. For example, infants quickly learn the statistical regularities in speech sounds (phonemes) and word patterns as they acquire language. The brain constantly predicts the next sound (like speech or music) based on previous experience. For instance, when listening to a sentence, the brain expects the next word based on grammatical rules and contextual meaning, which helps reduce mental effort by filling in gaps or clarifying ambiguities in sound or music.

The SLPC model described above is based on extensive research studies. Although the complexity of biological organisms makes it challenging to quantify the relevant parameters, the cause-and-effect relationships established through logical inference in design experiments are strong and widely accepted. However, I remain concerned that they are viewed as almost axiomatic and the foundation of neuroscience. In particular, when applied to our neural functions, SLPC reflects phenomenological interpretations of neural activities without explaining why the phenomena behave as they do.

To understand what drives SLPC in music, we can either adopt a black-box approach based on phenomenology, using it as a starting point for subsequent neural behaviors, or we can explore the fundamental principles behind SLPC for greater intuitive clarity. In this book, we will take a hybrid approach, combining what we know about SLPC so far with one of physics' most basic universal laws.

This chapter introduces the main principle that governs the phenomena of SLPC functions in music.

Energy and Thermodynamics

Although the SLPC concept in music is rooted in neuroscience, it also has natural connections and similarities to other fields like physics and engineering.

One can imagine the musical SLPC as an engineering process aiming to achieve a consistent outcome. The sequential events in the SLPC process begin with learning the music from experience and building a model in our brain for future use, helping predict what will happen next. When the brain's internal model of a musical piece matches the upcoming music, neural responses remain stable. The listener feels content, and the brain continues functioning with minimal effort. However, if the incoming music differs from expectations, the neural system signals an alert and activates related brain regions. This alert prompts the model to update its prediction, decreasing the likelihood of triggering the alert again next time. A predictive error is created to improve brain function, prepare us for hearing and reward-related events, and reduce the effort needed to enjoy musical activities in the future.

Generating predictive error signals and our ability to detect them require a lot of energy. Why does the neural system bother with this chain of actions, consuming the precious energy in our brains? For instance, to recognize a slightly different rhythm from the ongoing one, an event-related potential (ERP) can be identified and measured as a mismatch negativity (MMN) signal, recorded by EEG equipment. It typically takes about a microjoule (10^{-6} J) to produce a neural impulse, and thousands to millions of impulses are needed for the EEG to detect it. This process amounts to roughly a few tenths of a joule (0.1 J). That's a significant amount of energy, especially considering that the brain uses around 12 joules per second (12W) during moderate activity. Listening intently to music for an hour can consume up to a few joules of energy if the music has a few pleasant surprises every second. In addition, each ERP event can trigger numerous neural activities along the brain's hierarchy, activating other regions and increasing overall energy use.

Among the many tasks our brain constantly performs, it is strongly motivated to conserve energy. What better reason for the brain to seek and remain in the lowest possible energy state, so it doesn't waste energy unnecessarily?

What motivates this urge to move toward and remain at the lowest energy state? As it turns out, nature has the ultimate motivation. It tends to drive self-contained systems, like a nuclear reactor or the brain, toward and keep them in the lowest energy state, as dictated by the second law of thermodynamics (hereafter referred to as SLTD). This law, which underpins nearly everything involving heat and energy exchanges, has several appealing qualities: (1) it is universally applicable, (2) it encompasses any natural phenomenon, (3) it is simple and understandable in plain terms, and (4) it can be expressed mathematically.

Let's take a brief, one-page digression into thermodynamics and its universality before we apply that to the neuroscience of music.

The Laws of Thermodynamics

The development of thermodynamics has given us a framework for understanding energy, heat, and work. This framework is defined by the three fundamental laws of thermodynamics, which explain the principles that control energy transformations in physical systems. These laws form the basis of modern physics and engineering, offering important insights into how matter and energy behave in the universe. Each of the three laws of thermodynamics focuses on a different aspect of energy.

The first law of thermodynamics, also known as the law of conservation of energy, states that energy cannot be created or destroyed but only changed from one form to another. Since heat is involved in at least one of these forms, the term thermodynamics is used.

The third law of thermodynamics states that the entropy (a measure of the orderliness of a physical system; lower entropy implies higher orderliness or lower randomness; see the discussion for the second law below) of a system approaches a minimum as the temperature approaches absolute zero degrees (0 Kelvin). In other words, its entropy decreases as a system is cooled toward absolute zero.

The first and third laws are intuitively and verifiably true. However, they describe the static properties of energy and heat, but do not offer a dynamic description of how heat and energy behave and interact with their environments. The second law explicitly explains the direction of energy flow.

The second law of thermodynamics, SLTD (Entropy and the Direction of Natural Processes), states that systems always move toward equilibrium at the lowest energy position. Therefore, the system has no net energy flow at equilibrium and is in its lowest energy state because no further energy transformations can occur without external interference. However, multiple local lowest energy states may exist in different environmental contexts, and natural or artificial processes can happen to release excess energy and use it for various engineering purposes.

SLTD plays an important role in explaining why systems tend to reach their "lowest energy states." These are states where the system is most stable, usually marked by a high level of orderliness. It also introduces the idea of the "arrow of time" through the concept of entropy. Since entropy always increases over time in an isolated system, the second law helps us tell the difference between the past and the future. That's why processes like mixing coffee and cream happen naturally, but the reverse—such as the spontaneous separation of coffee and cream—does not. (Although entropy is implicitly part of the second law, it is a complex and subtle idea that this book does not explore in detail.)

SLTD and the Lowest Energy States

SLTD is universally true. The following examples of dynamic systems show that this law is inviolable and should apply to any system, including biological ones. They include naturally occurring systems and systems we manipulate that tend to evolve toward their lowest states while extracting energy from them.

Natural events following the second law include chemical reactions like those that happen when burning wood. It starts with the breakdown of cellulose and lignin in the wood, followed by combustion that produces heat and light. Afterward, the system of wood, air (oxygen), and remaining material is in a lower energy state than before the burn, releasing the energy stored in the wood.

The systems we use to produce energy range from ancient water-powered mills to modern hydropower plants. These systems include water, plants, turbines, and energy collectors like mills or electrical generators. Water-powered mills and hydropower plants generate energy from their initial states and transition to lower energy states.

According to the same second law, laser systems used in the photolithography process of integrated circuit manufacturing demonstrate how we manipulate materials to release optical energy.

Nuclear power plants manipulate the fundamental nuclear system (a single nucleus and its agitator: a slow neutron) to release stored energy, resulting in a much lower energy state (with two split nuclei), also consistent with the second law.

Suffice it to say, it is widely accepted that the second law dictates that any closed system progresses toward its lowest energy state, whether we let that happen or induce that to take place. Next, we will link the second law to music through the SLPC processes.

SLPC and SLTD In Music

The neurological processing of music reflects the connection between SLPC and SLTD. Neural signals start with converting music into auditory impulses, travel through auditory pathways, and end with feelings of pleasure or displeasure when someone likes or dislikes the music. This sequence of events is the main focus of Chapters 6 through 8.

After music is converted into electrical impulses, it travels through the auditory pathway on its way to higher levels of cognition via the brain's ascending hierarchy, passing through key brain regions like the auditory cortex before reaching the reward system. If the music is familiar to listeners, who have probably heard it several times before, either consciously or unconsciously, the process of statistical learning (SL) has built an internal model or memory of that music, capturing either the entire piece or its essential parts. The neural system is now in a comfortable state and responds to the music's fluctuations based on previous experiences. It's important to note that the memory includes not only the music itself but also neural responses, such as how emotions are linked with rewards, although these are stored in different brain areas. The internal model of the familiar music acts as a guide for the predictive coding process, helping the brain anticipate what's coming next.

Predictive coding (PC) also influences the entire journey of this music. When the same piece is played in its most common form, predictive coding forecasts its progression based on stored memory, and no predictive error message occurs if the incoming music matches expectations. However, when the music varies—such as being extra forte at certain points or differing in arrangement for different performers or conductors—a neural response called event-related potential (ERP) arises within the neural system. This produces a predictive error message that signals the change, traveling through a slower neural pathway to the early neural level (feedback/feedforward), which then updates the existing model. Alternatively, it enables the lower level to anticipate changes if the error indicates a deviation from the high-level plan, a process known as feeding the message forward.

Throughout the series of events, the neural system stays in a comfortable, low-energy state and prefers to remain there if the music continues to sound familiar, as heard before. When the music plays as expected, no ERP is produced, and the brain doesn't need to use extra energy, as previously explained. If not regulated, the neural system would change its state in response to altered music and use energy, raising itself to a higher energy state. The feedback and feedforward processes, discussed in the next section, help keep the system in the least disturbed, lowest-energy state, according to the SLTD.

Engineering Modeling Of Slpc Process

As a physicist-turned-engineer, I often use engineering models to explain the appearance and behavior of natural phenomena. Modeling the SLPC neural process for music as a general engineering control system with SLTD as the ultimate engineering goal may make SLPC's inner workings easier to understand in music.

Control System

A control system is a dynamic mechanism that manages input from various sources and regulates the expected output. The input could be an instigator, instruction, or music stream that the "system" processes to achieve a desired result. A control system might be an automobile, a set of instructions for a factory, or our neural system. The output of these systems could relate to the intended speed, the factory's production level, or the energy status of our neural system.

Control systems are found in almost every part of modern life. I'm sure we all know about cruise control in cars, which is a standard feature in most models today. In an older car, the vehicle's speed depends on input, specifically the throttle, which works through the engine and its connected mechanical parts to move the vehicle. When road conditions change, such as going uphill or downhill, the car will slow down or speed up if the throttle isn't adjusted accordingly. However, if a fixed speed is desired, the driver usually controls the speed by pressing the accelerator harder or lighter for throttle control. This interaction between the car and the driver shows a control system where the human is the controller.

When cruise control in modern cars is activated and set to a specific desired speed, the vehicle detects the difference between the actual speed and the set speed. It adjusts the throttle to accelerate or decelerate in proportion to the difference between the actual and the set speeds. This system is said to operate in a closed-loop control with FEEDBACK, so named because the input (throttle) and output (speed) are connected, forming a loop. This closed-loop allows for continuous adjustments, whether through hardware or software, until the speed matches the set target. Feedback plays a key role by relaying the speed back to the throttle to prompt appropriate action.

Alternatively, if a roadmap is pre-loaded into the journey based on previous trip memories, the vehicle can detect, via GPS positioning, for example, that changes are approaching; the throttle can then be adjusted to keep the car at its set speed. This system forms the basis of a closed-loop FEEDFORWARD control system. The timing of the feedforward control relative to the location data may be immediate or even anticipatory. However, this feedforward control system might be ineffective if the roadmap is unclear due to poor metrology or corrupted memory.

The terminology of "FEEDBACK" and "FEEDFORWARD" is often used loosely and sometimes incorrectly in some neuroscience interpretations of neural signals. We need to clarify the difference and use them properly for the SLPC, especially in the context of music. It is important to note that the dominant interactions in musical SLPC are of the feedforward type because the "learning" has created a working model in our memory.

The clearest difference between feedback and feedforward control can be summarized in one sentence: when actions occur before an event in a control system, they are called feedforward; when they happen after the event, they are called feedback. In other words, feedback control is a passive approach that waits for the outcome to determine if action is needed. On the other hand, feedforward control takes a proactive approach, predicting the required action and adjusting to keep results near the target. The predictive error in music perception and cognition (Chapters 6 through 8) is akin to the error between the map (expected) and the actual (GPS-determined) locations of the car in the cruise control analogy.

The benefits of feedback control systems include precise error correction and stability. However, they also tend to have slower response times because the system reacts only after detecting an error, which can cause delays in correction. Conversely, feedforward control systems offer faster responses and greater efficiency in predictable environments. If an unforeseen disturbance occurs, the system will provide instructions on how to counteract its effects in advance. Of course, this feedforward control also relies on accurate roadmaps.

There is, however, a situation where the feedforward process may not handle the cruise control example effectively. Assuming the stored roadmap is correct but the road is under construction with a temporary detour, the roadmap, unaware of the detour, provides a conflicting action plan to the controller, resulting in a failed feedforward control. The detour information needs to be sent to update the roadmap so the next trip will be correctly controlled. Such an adjustment can be made offline and doesn't need to be immediate, but it is necessary. Although not strictly classified as engineering feedback or feedforward, this process benefits the control system in the long run. Therefore, this type of control system can be considered an OFFLINE FEEDFORWARD closed-loop system. This situation also occurs quite often in music perception and cognition.

Since neither feedback nor feedforward alone is perfect, combining both approaches for tighter, faster, and more precise control seems reasonable. In addition, offline feedforward may add security if the process in a control system is to be repeated. Most control systems use both feedback and feedforward simultaneously, including our neural system during SLPC.

A typical example of this combined control is a rocket's trajectory, where the guidance system uses pre-calculated thrust adjustments based on the planned flight path (feedforward) while also monitoring real-time altitude and velocity data to make fine-tuning adjustments through thruster control (feedback). This ensures the rocket stays on course despite external disturbances like wind or atmospheric changes. However, if the launch site is near a mountainside with predictable air circulation, the control system would need to adjust the navigation system so that the next launch can be better controlled using offline feedforward control.

SLPC and Control System

It is most helpful to describe the closed-loop control system graphically. Figure 3.1 shows a combined feedback, feedforward, and offline feedforward control system that ensures both immediate and long-term operation integrity.

The main line in the center of the diagram connects the output to the input. In the cruise control example, the input is the throttle, and the output is the desired speed of the moving vehicle. The sensor is the speedometer, where the difference or error signal, between the actual speed and the desired speed, is sent through the controller to adjust the throttle based on the size of the error signal.

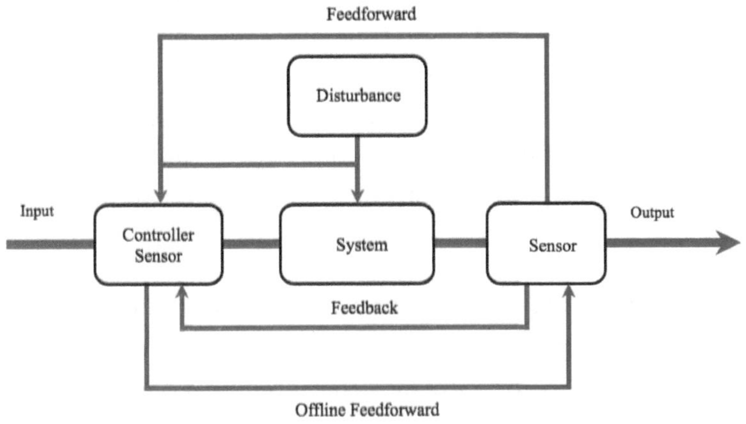

Figure 4.1. Illustration of a combined feedback, feedforward, and offline feedforward control system, including the immediate and long-term integrity of the operation.

What are the corresponding components in a musical cognition SLPC neural process? The parameters are more challenging to determine and characterize in a biological system. Nonetheless, we can view the input to this neural system as the music we hear. Statistical learning (SL) retains the music in our memory, similar to a pre-programmed roadmap for a feedforward cruise control system. The act of storing information in, for example, the hippocampus region, is also part of predictive coding (PC). The neural system operates continuously, converting music from acoustic signals into neural impulses that travel through the fast auditory pathways, interpreting the music in both temporal and spectral domains, and interacting with the reward system to support emotional interpretation. No predictive error is generated when the auditory neural processes detect nothing unusual relative to the stored model. The output of the neural system, expected to stay in a low energy state, hums along as various parts of the brain perform their respective functions.

However, event-related potentials (ERPs) are generated when music deviates from memory, signaling a predictive error (PE) to alert the neural system about the changes. The ERP originates from a comprehensive sensor that detects auditory or musical abnormalities needing attention. At the same time, PE passes onto the controller, closing the loop, to readjust the system's input (the actual input) and keep the output in a lower energy state. As we will see later, the neural pathways of the musical streams are optimized for different purposes and speeds to enable smooth feedback and feedforward control.

The upper part of the diagram shows the feedforward process. In the cruise control example, detecting adverse conditions such as disturbances must be fed forward into the input to keep the output on track. In music, the current mental model of a familiar piece is already stored in memory through statistical learning and can be seen as the roadmap in the cruise control analogy. When a crescendo is anticipated, the feedforward function kicks in, warning the hearing process about what is approaching, which triggers relevant neural responses, like a pleasant reaction, that occur the last time you heard this music.

You might have heard about an intriguing phenomenon called "missing fundamentals" in the study of musical cognition. Under certain testing conditions, people often perceive the fundamentals of a single-syllabic tone that actually lacks these fundamentals. This phenomenon occurs because of the musical feedforward control system in action. Our neural system's memory has learned, through SLPC, that every single-syllabic tone consists of multiple harmonics, with the lowest frequency being the fundamental component. It has been shown that a specific area in the auditory cortex is especially sensitive to tones that include both the fundamental and harmonic components, but it cannot distinguish between the different harmonics. As a result, even an artificial tone composed only of the second, third, and higher harmonics can produce the same perceptual response. Our brain cannot tell the difference and perceives the tone as having the fundamental component.

In the control system analysis, our mental image (i.e., the memory) of tones with the fundamental components is the norm, similar to the roadmap in the cruise control example. Upon hearing the tones without the fundamentals, the predictive error has no reason to alter the input to our cognitive neural system since it is still perceived as a tone with the fundamentals. Interestingly, this phenomenon also explains why humans intuitively recognize the existence of octaves in music, which forms the basis for the musical scales elaborated in the last chapter.

The lower part of the control system diagram also shows the offline feedforward path. In the cruise control example, this offline feedforward loop corrects the stored map after a change is detected, ensuring the next trip uses the updated map. The correction isn't needed immediately and can be done "offline." In an SLPC model, this path allows for offline encoding after detecting variation in familiar music and updates the memory for future use.

This comparison between an engineering control system based on fundamental physical laws and music's SLPC will be highlighted when necessary in the following chapters.

Summary

This chapter introduces high-level concepts of statistical learning and predictive coding (SLPC), which have quickly become fundamental to cognitive neuroscience over the past twenty years. This framework has been successfully applied to nearly all human senses: vision, olfaction, taste, touch, and, of course, auditory responses. The primary focus has been on the SLPC of music-related perception and cognition, which is a key aspect of auditory neural responses.

We then introduce one of the most essential laws of physics, which has remained unbreakable over the past three hundred years: the second law of thermodynamics (SLTD). This law governs spontaneous phenomena and engineering designs like nuclear reactors and lasers. More importantly, it influences how we perceive, learn from, and enjoy music.

Using an engineering control system as a model, we draw a parallel between that and the operation of SLPC in our neural system. We demonstrate that engaging in music naturally activates the SLPC while allowing the neural system to use the least amount of energy possible for brain function.

With the background established in this chapter, we can think, although not yet in precise or numerical terms, that our reactions to music stem from a natural tendency to bring our neural state to its lowest energy level before the musical stimulus reaches our auditory sensors. Along the way, there will be ERPs and predictive errors. Meanwhile, the neural system aims to minimize energy expenditure in achieving its goals.

As we navigate the complexities of academic research and the results related to music sciences, it is essential to recognize that a fundamental principle underlies the intricate processes occurring through the multiple phases connected by nested feedback and feedforward loops, as detailed in Chapter 1. The complexity stems from the many parameters needed to model SLPC under this core principle. However, nature will continually cause this biological neural system to settle into the lowest energy state, which forms the foundation of the science of music.

Bibliography

1. "The free-energy principle: A unified brain theory?" Friston,

K., *Nature Reviews Neuroscience,* **11**(2), 127–138 (2010). https://doi.org/10.1038/nrn2787

2. "Auditory expectation: The information dynamics of music perception and cognition," Pearce, M. T., & Wiggins, G. A. *Topics in Cognitive Science,* **4**(4), 625–652 (2012). https://doi.org/10.1111/j.1756-8765.2012.01214.x

Chapter 5

How We Hear

We are now ready to dive into the details of the train of music traversing our brain in the order of the phased-naturalism formulism (Figure 5.1, reproduced from Figure 1.1). The first phase involves purely physical presence, whether the music is heard by animals or humans, and does not include hearing itself. The second phase, which is the focus of this chapter, examines the auditory physiological processes that occur when music reaches our ears and is converted into neural impulses that the brain interprets and processes, preparing for further understanding.

Musical Phased-Naturalism

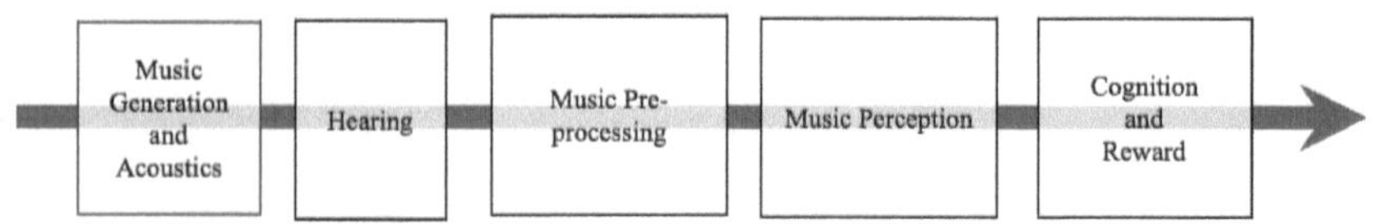

Figure 5.1. Same as Figure 1.1. The graphical representation of the phased-naturalism formulation.

This sequence, from the mechanical reception of sound vibrations to their conversion into neural impulses that our brain can interpret, is well-documented and likely familiar to some readers. However, this chapter approaches the topic differently by explaining the functions of spectrally specific sound-to-neural impulse conversion. Specifically, it describes how sound undergoes spectral analysis, an engineering process that separates sound into its individual frequency components. Figure 3.1 illustrates the outcome of this process of a human voice singing the C3 tone. Aside from this engineering perspective on the hearing mechanism, readers who are already familiar with the topic can proceed to Chapter 6, where we explore how music is processed in the brain's central auditory regions. As such, the description of this phase in this chapter will be brief.

The Hearing

Figure 5.2 shows the step-by-step process of hearing, illustrating how the ear converts sound waves into neural signals. Each step matches the function of a specific anatomical structure.

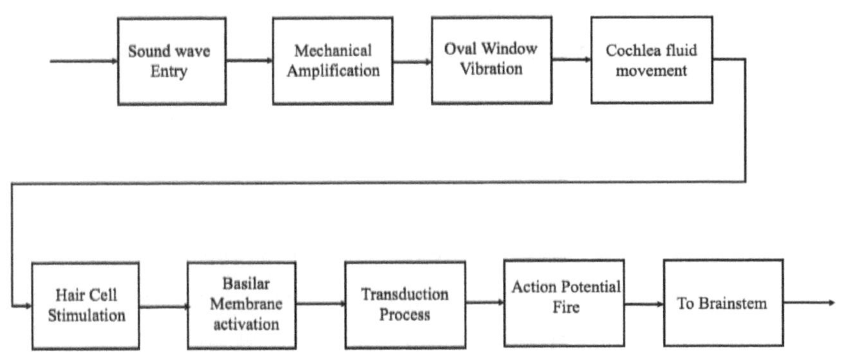

Figure 5.2. Illustration of the hearing process in a step-by-step chain of events, describing how the ear transforms acoustic disturbances into neural impulses.

Although the hearing process produces electrical impulses at this stage, they contain all the important information encoded in the sound, such as loudness, timing, rhythms, melodies, timbres, emotion, lyrics, and messages.

The first box directs and concentrates sound waves toward the ear canal, causing them to enter a narrow passage that ends at the oval window. The vibrating eardrum transfers its vibrations to the three tiny bones in the middle ear (ossicles: malleus, incus, stapes), amplifying these mechanical vibrations. The third box connects the stapes bone to the oval window, causing it to vibrate in response to the movement of the stapes. The fourth step generates pressure waves through the eardrum into the fluid-filled cochlea. The fluid inside the cochlea is set into motion by these pressure changes, creating a traveling wave that varies in intensity along its length. The sixth box stimulates the motion of rows of hair cells on the basilar membrane. The seventh box executes the transduction process, where fluid movement deflects the stereocilia (hair-like projections) on the hair cells. This mechanical distortion opens ion channels, causing ions to flow inward and fire up the action potential, or electrical impulses, which are then transmitted to the auditory nerve, as shown in the eighth box. Finally, the ninth step involves the auditory nerve carrying these electrical signals to the brain's auditory cortex, where they are processed and interpreted.

The Ear

The initial stage of hearing happens when sound physically enters our ears. The following briefly describes the three main parts and their functions: the outer, middle, and inner ear, separated by the membranes of the tympanic membrane and oval window, as shown in Figure 5.3. Each part has a specific role, which we will now examine.

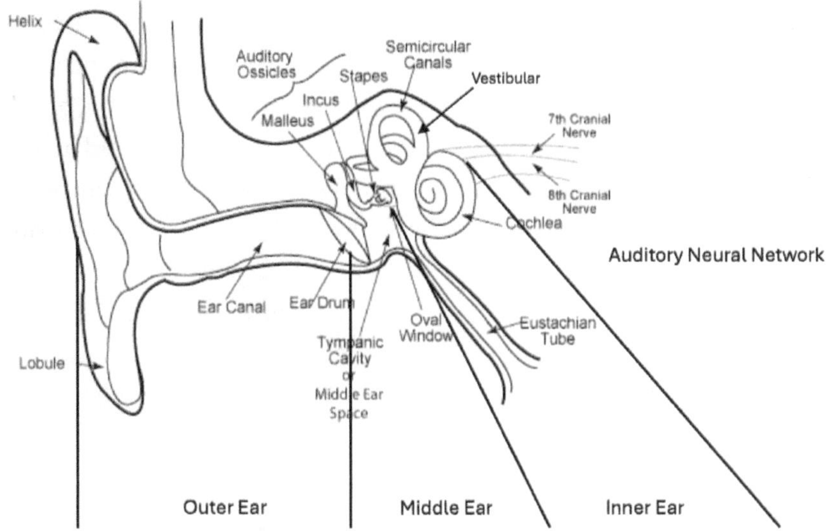

Figure 5.3. The ear structure consisting of the outer, middle, and inner ear parts.

Outer Ear

The outer ear consists of two main structures: the pinna (auricle) and the ear canal (external auditory meatus), which ends at the tympanic membrane that separates the outer ear from the middle ear.

Pinna (Auricle): The pinna, also known as the auricle, is the visible, cartilaginous part of the outer ear. Its complex shape and structure serve multiple functions. Primarily, the pinna collects sound waves from the environment and channels them into the ear canal. Its unique folds and contours assist in locating the source of sounds, helping us determine the direction they come from. It also reduces echoes bouncing in our surroundings. Additionally, the pinna amplifies certain sound frequencies, especially those in the range of human speech, improving our ability to communicate effectively.

Ear Canal (External Auditory Meatus): The ear canal, also called the external auditory meatus, is a tube that extends from the outer ear to the eardrum of the middle ear (tympanic membrane). Its main function is to transmit sound waves from the pinna to the eardrum. Lined with specialized skin cells, the ear canal produces cerumen, commonly known as earwax. Earwax serves as a protective barrier, helping to block foreign particles, dust, and insects from entering the ear canal. It also has antimicrobial properties that prevent the growth of harmful bacteria and fungi. Additionally, the shape and length of the ear canal help concentrate air pressure changes that impact the eardrum, or the tympanic membrane.

<u>Middle Ear</u>

The middle ear is a small, air-filled cavity located behind the eardrum. It contains three main structures: the eardrum (tympanic membrane), the ossicles (malleus, incus, and stapes), and the Eustachian tube.

Eardrum (Tympanic Membrane): The eardrum is a thin, semi-transparent layer that separates the outer ear from the middle ear. Its main role is to turn sound waves collected by the pinna and transmitted through the ear canal into mechanical vibrations. These vibrations are then passed to the ossicles, starting the process of hearing.

Ossicles (Malleus, Incus, Stapes): The ossicles are three tiny bones located within the middle ear cavity: the malleus (hammer), which connects to the tympanic membrane; the incus (anvil); and the stapes (stirrup), which connects to the oval window. These small bones form a chain-like, folded structure that connects the eardrum to the inner ear, providing a mechanical advantage like that of a crowbar. When sound waves hit the eardrum, they cause it to vibrate, transmitting these vibrations to the ossicles. The ossicles, in turn, amplify the vibration and transfer it to the oval window. This amplification is aided by the lever structure, allowing for efficient transmission of sound energy from the relatively large surface area of the eardrum to the smaller surface area of the oval window, resulting in greater displacement and maximizing the sensitivity and responsiveness of the oval window.

Eustachian Tube: The Eustachian tube does not have a role in hearing. It is a narrow passage connecting the middle ear to the back of the nose and throat (nasopharynx). Its main function is to equalize pressure between the middle ear and the outside environment, which helps prevent pressure buildup in the middle ear cavity. This pressure regulation is essential for maintaining proper hearing and avoiding discomfort or damage to the eardrum and other middle ear structures. The Eustachian tube also serves as a drainage pathway, allowing excess fluid and mucus to drain from the middle ear into the throat, helping to prevent infections and keep the middle ear healthy.

Inner Ear

The inner ear is located inside the temporal bone of the skull. It consists of two primary parts: the vestibular system and the cochlea.

The Vestibular System

The vestibular system doesn't directly assist with hearing. Instead, it manages balance, spatial awareness, and coordinates head and eye movements. It is made up of three semicircular canals and the otolithic organs (utricle and saccule), which are fluid-filled structures in the inner ear that are connected.

The semicircular canals detect rotational movements of the head in three-dimensional space, while the otolithic organs sense linear acceleration and changes in head position relative to gravity.

Within the otolithic organs, specialized sensory cells called hair cells detect the movement of tiny calcium carbonate crystals known as otoliths, which shift in response to motion. This movement bends the hair cells, sending signals through the vestibular nerve to the brainstem and cerebellum. These signals are then integrated with visual and proprioceptive information to help maintain balance and coordination. Dysfunction of the vestibular system can cause symptoms like vertigo, dizziness, and loss of balance.

The Cochlea

The cochlea (Figure 5.4) is the primary auditory organ, a spiral-shaped, fluid-filled structure that converts sound into neural signals. It starts at the oval window, which is connected to the stapes bone. This structure contains the sensory organs for hearing, called hair cells, which change mechanical vibrations into electrical impulses to be sent to different parts of the brain.

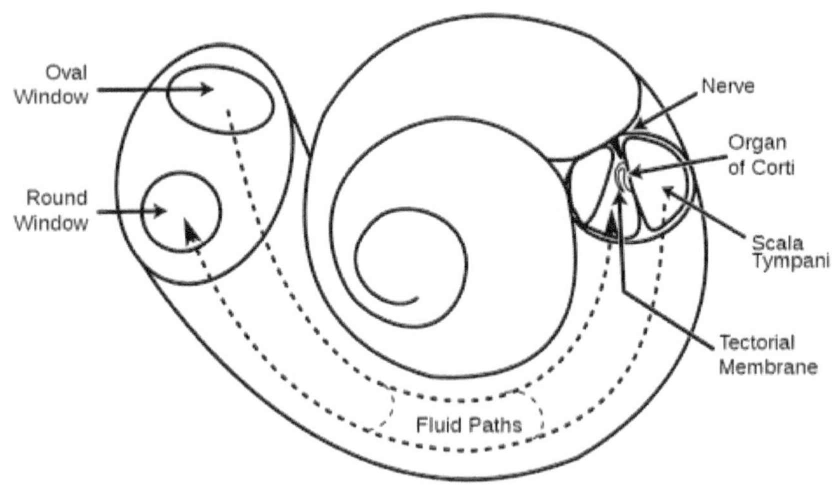

Figure 5.4. The cochlear structure.

If uncoiled, the cochlea would resemble a long, cone-shaped tube that extends from a wider end, called the base, to a narrower end, known as the apex. It measures approximately 35 mm in length with an internal diameter of 1.6 to 2.6 mm. Inside the cochlear tube are three fluid-filled, membranous chambers called sub-tubes, arranged parallel to each other: the scala vestibuli (SV) or vestibular duct, the scala media (SM) or cochlear duct, and the scala tympani (ST) or tympanic duct. The SV and ST contain perilymph, while the SM contains endolymph. Physical hearing takes place within the scala media.

Figure 5.5 illustrates how the cochlea functions, with the oval window serving as the initial contact point for mechanical movement. Sound waves, which involve compression and rarefaction of air, are transmitted to the fluid inside the cochlea's scala media sub-tube through the oval window.

The cochlea detects sound through both the temporal aspect, which follows the waveform, and the frequency aspect, which uses location to distinguish different pitches. Therefore, the sensors along the cochlea perform these two functions. Figure 5.5 shows the main parts: (1) the three membranous, fluid-filled chambers that run parallel—scala vestibuli (SV) or vestibular duct, scala media (SM) or cochlear duct, and scala tympani (ST) or tympanic duct; (2) the organ of Corti, supported by the basilar membrane, which contains the hair cells, hair, and tectorial membrane; and (3) the sensory nerves and their bundles.

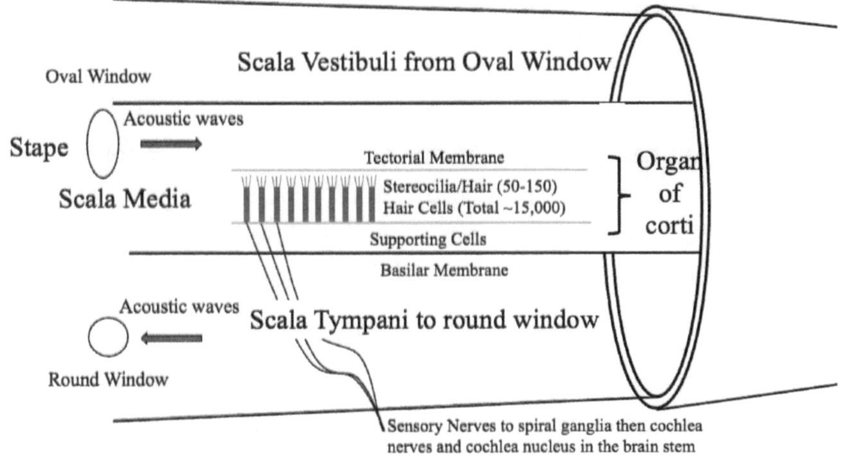

Figure 5.5. Details of the key functional parts for neural impulse generation.

The organ of Corti is the center of the sensory system; it contains about 20,000 hair cells, each with a bundle of fibers called stereocilia or hair at its tip. The stereocilia are so sensitive that they can detect movement pushing them sideways by as little as the width of a single atom. To put this into perspective, the swaying of stereocilia, which equals the width of an atom, is comparable to the Eiffel Tower swaying half an inch in the wind.

Cochlea as a Frequency Detector

Let's focus on how this elongated, tubular, liquid-filled structure can detect pitches. It utilizes two complementary aspects of sound to do this: temporal and positional. First, the temporal aspect of sound is shown in the intensity of the pressure wave disturbances, which are represented by their waveforms. The pressure increases when the liquid is compressed and decreases during rarefaction. The second aspect is that any sound or tone contains more than one vibration frequency because the waveform of naturally produced sounds is rarely a pure, idealized sinusoidal vibration. Instead, it's always a mixture of an ideal wave and additional components at multiple harmonic intervals of the fundamental frequencies, as shown in Figure 3.1.

When sound travels through the liquid inside SM, its intensity weakens along the cochlea. Think of swimming in a pool; sounds from the surface are muffled. Additionally, you've likely noticed that high-pitched screams are dulled more than low-pitched voices. This occurs because sound is "selectively attenuated" by the water's natural properties, a common physical phenomenon.

Selectivity attenuation describes its unevenness across different acoustic frequencies; it is more prominent for higher audio frequencies than for lower ones. This attenuation increases proportionally to the square of the audio frequency. In other words, low-pitch sounds can travel much farther than high-pitch sounds along the cochlea. Since our hearing spans a frequency range of four orders of magnitude, acoustic attenuation can vary greatly from low to high frequencies, making the cochlea a highly sensitive frequency detector. Imagine a sensor for energy or vibrations at every point along the cochlear tube (Figure 5.6). These sensors detect higher frequency acoustic energy near the beginning of the cochlea, or base, and lower energy toward the tail end, or apex. This highly frequency-specific "attenuation" functions much like a well-designed spectrum analyzer, where the position of each sensor helps differentiate between various sound frequencies.

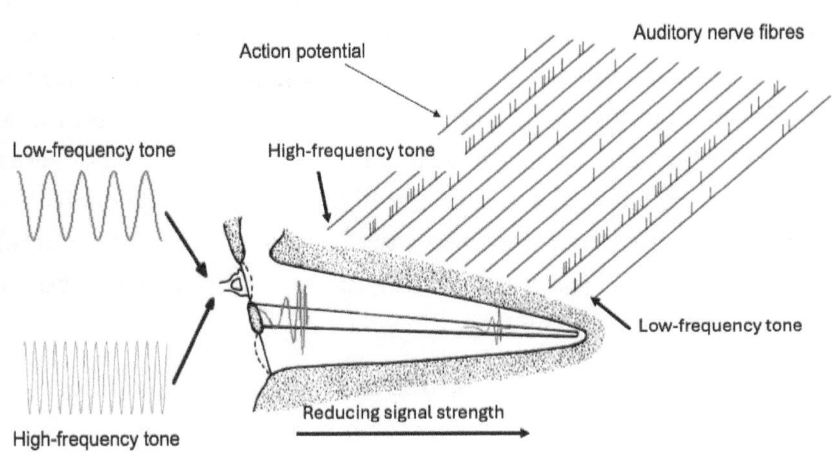

Figure 5.6. *The cochlea functioning as a spectrum analyzer.*

Taking advantage of the highly selective attenuation of sound frequency content, the organ of Corti acts as a frequency spectrum analyzer. Lower-frequency sounds produce impulses (explained in the next section) just as quickly as the pressure waves and reach the farthest part of the cochlea, the apex. In contrast, higher-frequency sounds activate hair cells near the base of the scala media chamber. For example, a tone of 600 Hz is translated into 600 neural impulses per second. This lower-pitch sound detection is thus called temporal coding.

To handle the higher speeds needed for high-pitched sounds, the stereocilia cannot respond as rapidly and must work together in a volley system where different nerves fire in sequence, enabling us to detect sounds beyond about 4,000 hertz. Therefore, the pitch of high sounds is partly determined by the location within the cochlea that generates the action potentials. This position-sensitive hearing of high-pitched sound is called position coding.

<u>Limitations of the Cochlea</u>

Since the organ of Corti is mainly mechanical and has inherent mechanical limits, it can accurately follow the precise movements of stimuli at slower speeds but has difficulty keeping up with higher speeds. The sensors for lower-pitch sounds respond to the temporal profile of vibrations, enabling temporal detection of sound at deeper parts of the cochlea. The sensors for higher-pitch sounds detect the aggregated vibrational energy near the front of the cochlea. In other words, like any laboratory spectrum analyzer, our cochlear resolution and sensitivity are limited.

The limitation prevents the cochlea from detecting every frequency found in nature. It is responsible for human perception of sounds from 20 Hz to 20,000 Hz, with a resolution of about 1/230 octave. (Remember that an octave contains 1200 cents, as discussed in Chapter 3; therefore, 1/230 of an octave is nearly impossible for most people to distinguish, being roughly 5 cents.) The cochlea's frequency range limitations are evident through frequency response curves, as shown in Figure 5.7. The blue line and band indicate the typical response for modern humans and for others.

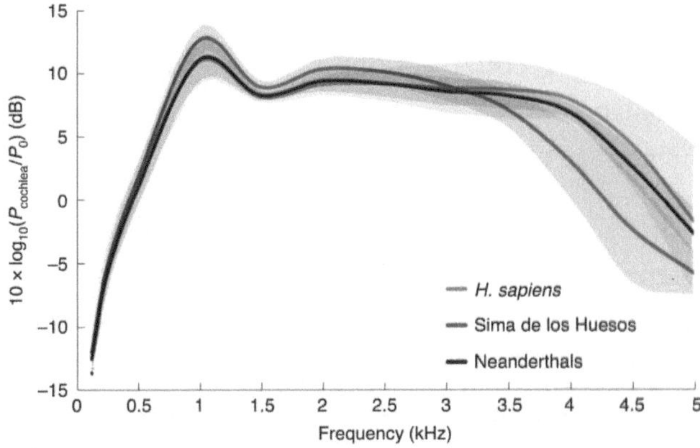

Figure 5.7. *The frequency responses of the cochlea in humans and our ancestors.*

This figure shows that for average human hearing, our cochlea can easily respond to frequencies below 4,500 Hz. The shapes of this response curve will vary between individuals and with aging, in particular at the higher frequency range. As such, our appreciation of music might be compromised for older people and the hearing-impaired.

On a side note, from this figure, we can answer some long-standing questions about how our musical abilities compare with those of our ancient ancestors. The responses of humans (Homo *sapiens*) hearing is plotted against Neanderthals (who went extinct about 40,000 years ago) and the Sim de los Huesos hominins (believed to be early Neanderthals), who lived 430,000 years ago, about 200 miles north of Madrid, Spain. Note that both the frequency and response are on logarithmic scales, indicating the wide ranges they cover. The frequency range spans at least 10 octaves, while the magnitude covers approximately a 25 dB (~400 times) range. The similarities between H. sapiens and Neanderthals are interesting. We mentioned in Chapter 3 that the earliest flutes were invented by Neanderthals around 50,000 years ago. This figure shows that modern humans' and Neanderthals' ear structures are similar, though not identical, indicating comparable hearing ability. A replica of the Neanderthal's flute, based on 3D reconstruction, plays the notes that we modern humans can also hear. This feat is not surprising because of the similarity of their spectral responses.

However, the audio response of ancient hominins, particularly at high frequencies, is about 3 dB (twice) less sensitive than ours around 4,200 Hz. This suggests that Neanderthals may not have been able to hear the C8 key on a piano as well as we can. Would the instruments they made produce the same high-pitched tones and serve the same musical functions as those we, H. *sapiens*, created?

As a side note, we can infer from this figure that it took hominins approximately 400,000 years to develop our current hearing abilities, particularly in the higher-frequency range, starting from the time of the Sima de los Huesos hominins. This inference is based on the assumptions that the Sima de los Huesos hominins are our direct ancestors, which we know they are not from morphological evidence. This speculation is merely a point of interest for discussion.

The Action Potentials

As shown in Figure 5.5, the tectorial membrane near the stereocilia causes the hairs to bend and release neurotransmitters, which trigger electrical impulses in the auditory nerve. This movement starts a chemical reaction that produces neural signals, which are then received by the auditory nerve fibers.

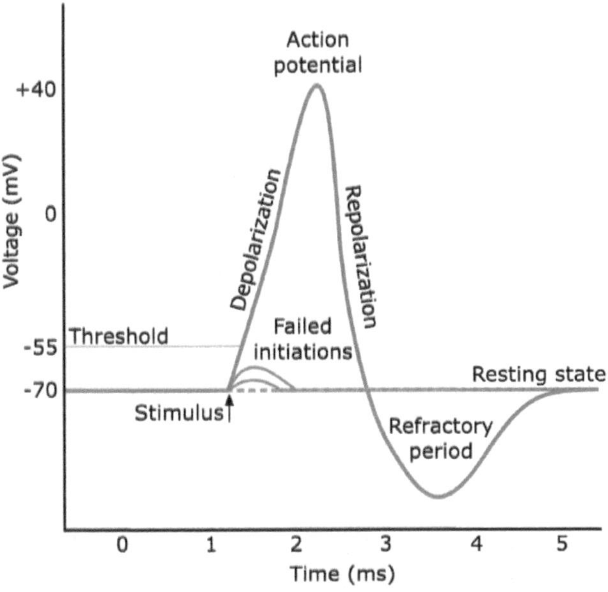

Figure 5.8. *The action potential pulse shape and its four phases: resting potential, depolarization, repolarization, and hyperpolarization.*

As mentioned earlier, when sound waves cause the fluid inside the inner ear to move, the stereocilia of the hair cells bend. This bending opens mechano-transduction channels at the base of the stereocilia. These channels let ions, especially potassium and calcium, flow into the hair cell. The influx of ions changes the membrane potential of the hair cell. Specifically, the entry of positively charged ions depolarizes the cell, causing a change in voltage across the cell membrane. This shift in membrane potential then triggers the release of neurotransmitters, mainly glutamate, from the base of the hair cell.

These neurotransmitters are released into the synaptic cleft, the small space between the hair cell and the afferent nerve fibers. The released neurotransmitters bind to receptors on the afferent nerve fibers, especially the dendrites of spiral ganglion neurons. This binding triggers the generation of action potentials in the nerve fibers. The action potentials then travel along the afferent nerve fibers to the central nervous system, specifically the auditory nerve for hearing or the vestibular nerve for balance. Finally, the neural impulses reach different areas of the brain, including the auditory cortex for hearing. Here, the brain processes these signals, allowing us to perceive and interpret them.

Experimental and theoretical studies have explored the properties of these neural impulses, known as action potentials. Since we will later discuss how these impulses travel through our brains, it is helpful to understand what they are. The following is a high-level overview of how they are generated and their features.

The action potential is a short electrical impulse that moves along a nerve fiber. This electrochemical process allows the transmission of information between neurons and target cells. The action potential waveform usually has several distinct phases: resting potential, depolarization, repolarization, and hyperpolarization. Figure 5.8 shows the shape of the impulses.

Resting Potential: When a neuron is at rest, it maintains a negative charge inside compared to the outside, usually around -70 millivolts (mV). This is mainly because of the uneven distribution of ions across the cell membrane, with a higher concentration of sodium (Na+) and chloride (Cl-) outside, and a higher concentration of potassium (K+) inside.

Depolarization: When a neuron receives a strong enough stimulus, voltage-gated sodium channels open on the membrane, allowing Na+ ions to rush into the cell. This influx of positive ions depolarizes the membrane, making the inside less negative. This rapid voltage change initiates the rising phase of the action potential.

Repolarization: When the membrane potential reaches its peak, the voltage-gated sodium channels close, and the voltage-gated potassium channels open. Potassium ions exit the cell, repolarizing the membrane and restoring its negative charge. This marks the falling phase of the action potential.

Hyperpolarization: In certain neurons, the efflux of potassium ions temporarily exceeds the resting potential, causing a brief hyperpolarization phase. During this period, the membrane potential becomes more negative than the resting state before returning to baseline.

Action Potentials Signals

The voltages for each phase of the action potential vary depending on the specific neuron and experimental conditions, as shown in Figure 5.8. However, typical values include:

Resting potential: -70 mV, as stated before.

Threshold for firing: The threshold for triggering an action potential is around -55 to -50 mV.

Peak Depolarization: The maximum voltage during depolarization ranges from about +40 to +50 mV. Voltage swings can be as large as approximately 100 mV. Each action potential firing consumes about one microjoule of energy (Chapter 4), which the brain would like to conserve if possible.

Repolarization: The membrane potential returns to its resting level, approximately -70 mV.

Hyperpolarization: The voltage can briefly fall below the resting potential, reaching as low as -80 mV.

These pulses exhibit large voltage swings and carry significant energy. Unlike the tiny voltage swing of about 7 μV in the air (which is a hundred thousand times weaker than these neural pulses), typically for an iPhone during a smooth conversation, this huge electrical voltage would almost certainly create disruptive noise in nearby nerves and in our brain (Chapter 9).

The rise and fall times of an action potential refer to the durations of the depolarization and repolarization phases, respectively. Several factors, including ion channel kinetics, membrane properties, and the strength of the stimulating signal, influence these timescales.

Rise Time: The rise time of an action potential is generally rapid, occurring within milliseconds, which reflects the swift influx of sodium ions during depolarization.

Fall Time: The fall time, which relates to repolarization, is slightly longer than the rise time because of the slower efflux of potassium ions and the inactivation of sodium channels. It usually lasts from a few milliseconds to several tens of milliseconds.

The action potentials carry sound information about (1) the changes in air pressure on the eardrums, (2) audio frequencies, and (3) loudness. At relatively low-pitched sounds, likely below 1,000 Hz, the neural impulses follow changes in air pressure, creating a one-to-one relationship between impulses and frequencies. As previously mentioned, the mechanics of the hair cells and stereocilia do not respond as quickly as the pressure waves that cause them. Instead, they collectively move the hair cell bundles and generate neural impulses, which do not necessarily match the frequencies of the rapid movements. These movements are produced by the hair cells, which inherently encode frequency information based on their position along the cochlea. However, it is incorrect that louder sounds cause higher voltage swings in the action potentials. Even louder sounds cause more frequent firing of neural impulses with a certain signature. However, there is no confusion about the additional impulses related to the audio frequencies, as they are already encoded in the positions of the hair cells.

Summary

This chapter focuses on hearing, especially the physical and physiological aspects. With the foundation we establish here, the next logical step is to trace these neural impulses to the higher levels of our brain to understand how music can eventually evoke our emotional responses.

Bibliography

1. "Neanderthals and Homo *sapiens* had similar auditory and speech capacities," Mercedes Conde-Valverde, Ignacio Martínez, Rolf M. Quam, Manuel Rosa, Alex D. Velez, Carlos Lorenzo, Pilar Jarabo, José María Bermúdez de Castro, Eudald Carbonell, and Juan Luis Arsuaga, *Nature Ecology and Evolution*, **5**, 609-615 (2021).
2. "TRPA1 is a candidate for the mechano-sensitive transduction channel of vertebrate hair cells," Corey, D. P., García-Añoveros, J., Holt, J. R., Kwan, K. Y., Lin, S.-Y., Vollrath, M. A., Amalfitano, A. Zhang, D.-S. *Nature*, **432**, 723–730, 2004.

Chapter 6

Music Preprocessing

The last chapter discusses the processes involved in the hearing phase of the phased-naturalism formulism. Hearing itself does not differentiate between types of sounds; these could be language, noise, or music. This chapter focuses on the beginning of music perception, called the music pre-processing phase. It breaks down the raw sound information of music into hierarchical piece parts for this and future phases to manage.

We explained how sound is transformed into neural impulses in Chapter 5. Now, we will examine what happens to this musical information as it travels through the auditory pathways to the brain's auditory center (auditory cortex).

After the music is transformed into electrical neural impulses, known as action potentials, they then travel through the brainstem and reach the auditory cortex, where music perception begins. These auditory pathways, which ascend and descend the neural hierarchy, are essential for helping the brain interpret the music we hear along the way. These pathways also use statistical learning and predictive coding to process and anticipate music, allowing us to recognize familiar patterns beyond the raw action potentials from the cochlear nucleus.

This preprocessing is to make sense of the music, essentially building our understanding by categorizing its elements, such as pitch, melody, rhythm, and harmony. Afterwards, the auditory cortex relays the information to higher brain regions, like the limbic and emotional areas, where it will be interpreted more deeply.

This chapter will also frame statistical learning and predictive coding (SLPC) as a control system model, focusing on music pre-processing.

Auditory Neural Impulses Transit

Once the ear detects an auditory stimulus, such as music, the neural impulses travel through the auditory pathways to the auditory cortex. The pathways involve multiple stages, but basically, they handle and categorize the raw data and leave the more complex perception processing for later.

The neurons in the auditory pathways can be broadly divided into two main types: lemniscus and non-lemniscus neurons. The lemniscus pathway plays a vital role in transmitting sensory information to higher brain centers, carrying precise signals quickly. They are essential for processing sound and vibration, enabling the brain to interpret these signals for complex auditory tasks. Lemniscus neurons mainly travel through the brainstem, including the medulla and pons, and eventually reach the thalamus before further processing in the auditory cortex. These neurons are thicker and heavily myelinated, giving them a light-colored sheath and making their axons better waveguides for action potentials. Better waveguides are less affected by surroundings and can transmit signals over longer distances. These neurons are the primary components of the white matter in the subcortical regions of the brain.

The non-lemniscus neurons transmit signals more slowly than lemniscus neurons because of their thinner myelin, giving them a more grayish appearance. These neurons are involved in passing less time-sensitive information, helping to refine the auditory process and provide context to incoming sounds, passing them along to higher brain levels. The signals from non-lemniscus neurons usually have shorter reach. Although their conduction speeds are slower than lemniscus neurons, they are essential for transmitting processed auditory signals between different parts of the brain, enabling it to assemble various pieces of information from sound over time.

Neural Impulses and Solitons

To better understand how auditory neural impulses travel along these pathways with precision and reach, it is helpful to draw an analogy to a concept in physics called *solitons*. Solitons are solitary waves in water channels by themselves. They are theorized as wave-like solutions to nonlinear wave equations that preserve their shape and speed over long distances. What makes solitons especially interesting is their ability to move without changing form, unlike typical waves that tend to lose their original characteristics. The solitons and auditory neural impulses have similar characteristics.

The concept of solitons was first discovered by John Scott Russell, who observed a solitary wave in the Union Canal in Scotland in 1834. He noticed a peculiar phenomenon: a single wave would travel down the canal without changing its shape, as though it were a solitary, persistent entity. In the following years, this phenomenon was theorized and reproduced in controlled settings. The solitary wave nature is given a pseudo-particle name: soliton, similar to how physicists name entities with a ballistic and particle nature. The suffix "-on" was adopted and applied to particles like the electron, proton, neutron, fermion, boson, lepton, hadron, and finally, solitons.

In the following years, solitons were extensively studied and generated interest across various fields of physics, from fluid dynamics to nonlinear optics. Jumping ahead to the 1970s, the concept of solitons was applied in a completely different area—fiber optics. The development of fiber optic communication transformed telecommunications, and a major breakthrough occurred when Hasegawa and Mollenauer, the latter a colleague and fellow physicist of mine, discovered the phenomenon of optical solitons. In a groundbreaking paper, Mollenauer and his colleagues showed that light pulses could travel through fiber optic cables under certain conditions without changing their shape, just like the solitary waves described by Korteweg de Vries et al. over a century earlier.

The discovery of optical solitons was a major theoretical breakthrough with practical applications. Mollenauer and his colleagues demonstrated that soliton propagation allows data to be transmitted over thousands of miles with minimal distortion, greatly boosting the capacity and efficiency of fiber optic communication systems. An unavoidable small amount of impurity in the optical fiber weakens the soliton during its long journey, requiring periodic replenishment to maintain travel over the entire distance. Notice the similarity between the neural impulse and the optical soliton, as shown in Figure 6.1, where the top part illustrates the shape of the action potential pulse we learned about in Chapter 5, and the bottom part displays that of a single soliton.

Although this technology is impressive, optical soliton transmission for optical communication has been replaced by more advanced technologies, especially with the rise of the DWDM (Dense Wavelength Division Multiplexing) technology and the powerful digital signal processing in fiber optic systems that span the globe and are housed within data centers. I was part of the massive effort that eventually obsoleted the soliton communication technologies.

Just as a soliton maintains its shape and speed, action potentials in neural fibers preserve their characteristic "all-or-none" property. When an action potential is generated, it occurs at full amplitude or not at all and propagates along its path. This consistency in the form and strength of the action potential ensures reliable signal transmission across neurons. This ensures that auditory information remains intact as it travels through the brainstem and thalamus, ultimately reaching the auditory cortex.

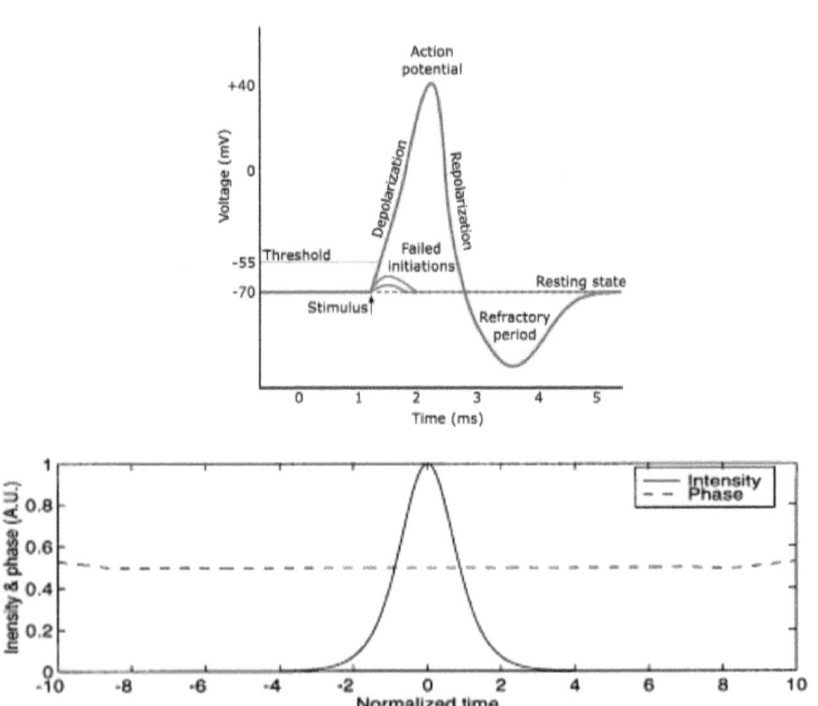

Figure 6.1. Comparison of the action potential pulse shape and that of a soliton.

Reaching Auditory Cortex (AC)

The auditory pathway facilitates communication between the cochlea and the auditory cortex. It uses the lemniscus neurons in the ascending pathways and non-lemniscus descending neurons in the descending pathways.

Hemisphere Lateralization

Our brain has two hemispheres, the right and the left. Both are activated while hearing music, but the auditory pathways are lateralized from the auditory nuclei, with either side specialized a little differently.

The Left Hemisphere: The left hemisphere is often called the "what" hemisphere. It mainly handles language and analytical tasks. This hemisphere dominates language understanding and production, including speech perception, word recognition, and syntax. In auditory processing, the left hemisphere is especially good at decoding and categorizing sound details, such as speech patterns, phonemes, their rhythm, and timing. The left hemisphere also processes music, but is oriented toward the music parts that are similar to the character of languages, such as the lyrics.

The Right Hemisphere: The right hemisphere, often called the "how" hemisphere, mainly handles spatial processing, understanding context, and perceiving environmental sounds. It is more sensitive to music, prosody (the rhythm, pitch, and intonation of speech), and nonverbal sounds. The right hemisphere is essential for combining auditory information with wider context clues, such as emotional tone or the location of sound sources in space. Additionally, it plays a key role in recognizing melodies, environmental noises, and contextual signals that affect the meaning of speech or music.

In sum, the auditory system has an inherent, specialized, and lateralized structure, with each hemisphere contributing to different aspects of auditory perception and processing. This division of labor is also evident within the AC, which divides the perception task into two layers: "what" and "how," as we will see later in this chapter. This lateralization leads to minor asymmetry between the physical topologies of the right and left hemispheres. Since we will mainly focus on music processing, the rest of the book will emphasize the right hemisphere while maintaining awareness of this asymmetry.

The Ascending Auditory Pathway

We start the discussion on the auditory pathway by displaying in Figure 6.2 a view of the right hemisphere with the auditory cortex (AC) shaded darker.

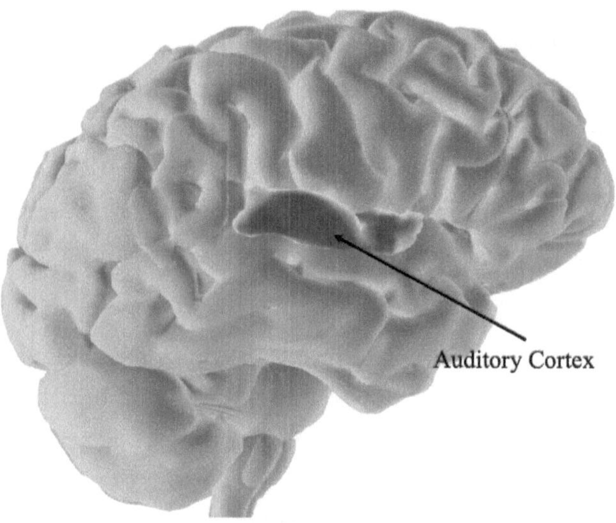

Figure 6.2. The right hemisphere and the location of the auditory cortex, which is shaded in a darker color. The brain outline is based on the 3D atlas from Neurotorium, https://neurotorium. org/tool/brain-atlas/.

In brief, the AC is located in the temporal lobe of the brain, specifically on the superior temporal gyrus (STG), within the Sylvian fissure. It is mainly hidden from view, buried deep within the lateral sulcus.

The auditory pathway is a hierarchically structured system through which sound information travels from the cochlea to the auditory cortex. This pathway involves both subcortical and cortical structures, with data being processed and refined at each stage. Figure 6.3 is a 3D graphic showing the approximate locations of the main parts of this neural pathway, viewed from the right back through the semi-transparent outline of the brain. The paths from the auditory nucleus to the auditory cortex, drawn as black lines, are visible in the diagram.

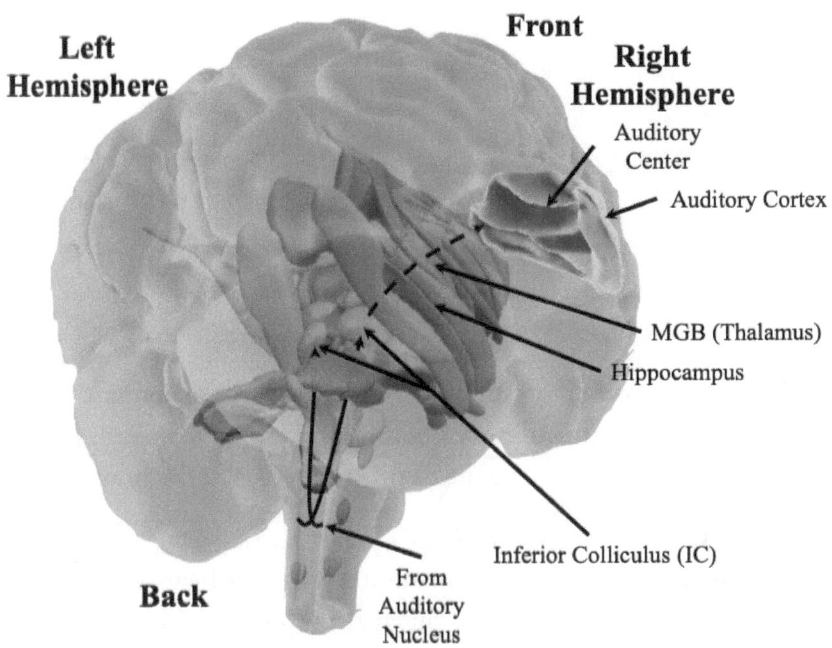

Figure 6.3. A 3D graphic indicating the approximate locations of the key parts of the neural pathway, viewed through the semitransparent outline of the brain from the back right direction. Drawn as black lines are the paths from the auditory nucleus to the auditory cortex. The brain outline is based on the 3D atlas from Neurotorium, https://neurotorium.org/tool/brain-atlas/.

Figure 6.4 follows the black line in Figure 6.3 and illustrates the logical flow of auditory signals through the pathways. The ascending auditory pathway can be divided into several key stages:

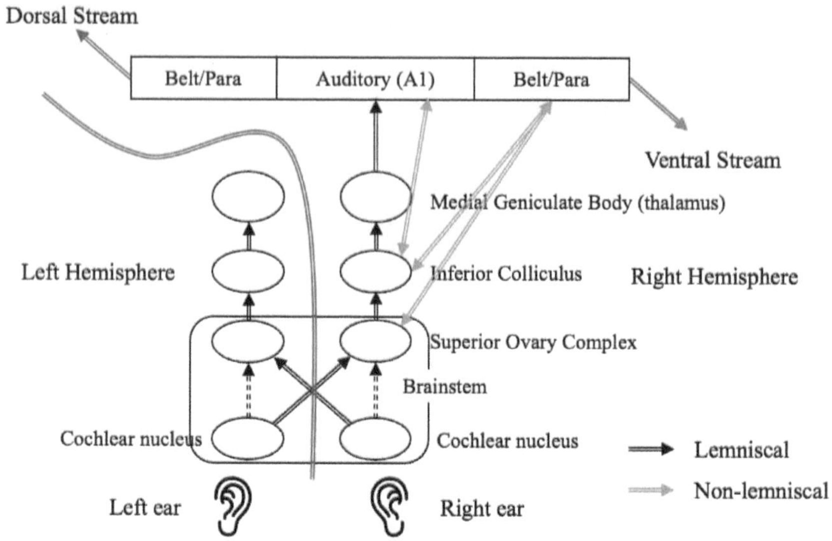

Figure 6.4. This figure illustrates the logical flow of auditory signals through the auditory pathway.

Cochlea: The nerve fibers that carry auditory action potentials are bundled together into auditory neurons and end in the spiral ganglia (Figure 5.4), then proceed to the cochlear nucleus in the brainstem.

Superior Olivary Complex (SOC) and Crossover: After initial processing in the brainstem, the auditory signals bearing the music information reach the superior olivary complex, which plays a vital role in sound localization. The SOC receives binaural input, meaning it processes signals from both ears. The crossover of auditory pathways occurs at this stage, with the left ear's auditory information traveling to the right hemisphere and vice versa (Figure 6.4). This decussation (crossing over) allows the brain to compare the timing and intensity of sounds received at both ears, aiding in the localization of sound sources in space. The importance of this crossover is that it enables both hemispheres to contribute to spatial auditory processing, regardless of which ear the sound originates from.

Inferior Colliculus (IC): From the SOC, auditory information ascends to the inferior colliculus (IC) in the midbrain. The IC is a central hub for integrating auditory signals, coordinating data from the cochlear nuclei and SOC. It also participates in reflexive responses to sound, such as turning the head toward a sound source. The IC further refines the auditory signal, preparing it for higher-level processing in the thalamus and cortex. Additionally, it combines input from various sensory modalities, like visual and tactile stimuli, enhancing spatial auditory processing.

Medial Geniculate Nucleus, or Body (MGN, MGB): The next step in the ascending auditory pathway is the medial geniculate nucleus of the thalamus (MGN). The MGN acts as the main relay station for auditory information before it reaches the auditory cortex. It further processes and enhances the auditory signals, dividing them into different frequency channels for detailed analysis (for tonotopy, see Figure 6.5). The MGN then sends this refined information to the auditory cortex, where it is processed at even higher levels of complexity.

Auditory Cortex (A1 and Belt Areas): The auditory signals reach the auditory cortex, specifically the primary auditory cortex (A1) and surrounding belt areas. The auditory cortex is responsible for higher-order processing of auditory information, such as sound recognition, speech processing, and music perception. It is divided into tonotopic regions, where different frequencies are processed in distinct areas. The core regions of the auditory cortex (A1) handle basic sound analysis, while the belt and parabelt regions contribute to more complex functions like auditory memory, attention, and perceiving sound context.

<u>The Descending Auditory Pathway</u>

While ascending auditory pathways are responsible for processing incoming sound and music information, descending pathways mainly provide feedforward (the term "feedback" is a misnomer in neural science and is clarified in Chapter 4) modulation to enhance auditory processing. However, feedback is also active through the same pathway. These pathways consist primarily of non-lemniscal neurons, which are generally slower and involved in carrying processed information. This processed information can travel down to lower brain regions, such as the brainstem, to influence how sound is processed at earlier stages. For example, predictive error signals can be sent downward to alter the perception of sound based on prior knowledge or expectations. This feedback/feedforward mechanism is essential to auditory perception, enabling the brain to "tune" to what one wants to hear. This tuning via predictive coding aligns with the second law of thermodynamics, suggesting that systems naturally aim to reduce surprise or error.

More specifically, this tuning allows listeners to "hear what they want to hear," thereby enhancing their perception of familiar sounds while filtering out irrelevant or unexpected stimuli. For example, in a noisy environment, the brain might "expect" the voice of a loved one and amplify that signal while suppressing others. This dynamic system of predictions and error correction forms the basis of much of our auditory perception, especially in complex settings like conversations or listening to music.

SLPC in the Auditory Pathway

Here's an overview of how statistical learning and predictive coding function along the auditory pathway, from the brainstem to the auditory cortex, specifically in music perception.

Statistical Learning in the Brainstem

Early structures in the brainstem, including the cochlear nucleus, superior olivary complex (SOC), and inferior colliculus, begin processing sound stimuli and extracting regularities from the auditory environment, such as temporal and spectral patterns.

At this stage, statistical learning already helps the auditory system recognize sound patterns, whether musical or environmental. For example, it enables the brainstem to detect repetitive sounds (e.g., rhythmic beats) or pitch contours (e.g., melody), which are common in musical stimuli. One can already see that sounds or music have two aspects, one involving time and the other involving frequency.

The brainstem can predict aspects of incoming sounds, increasing sensitivity to key auditory features. This process isn't as sophisticated as in the cortex, but it is essential for preparing the auditory system to process complex sounds further along the pathway. For example, it has been observed that the brainstem activates while distinguishing the consonance of intervals and chords (Chapter 9).

Predictive Coding in the Brainstem

In the brainstem, the auditory system begins to make basic predictions about the temporal structure of sounds, such as when a sound will happen (e.g., predicting the next beat or syllable). Brainstem structures like the inferior colliculus support this process, which helps analyze the sound's timing and frequency (pitch and melody) components.

The superior olivary complex (SOC), which processes binaural cues like timing and intensity differences between the ears, also contributes to predicting sound location. It aids the auditory system in anticipating where sounds are in space, which is vital for understanding speech and music. It is also crucial in locating where potential threats come from that could jeopardize survival.

The Interaction Between SL and PC

The auditory pathway from the brainstem to the auditory cortex is a highly adaptable system that effectively processes music using statistical learning and predictive coding. Statistical learning enables the auditory system to recognize patterns in sound, while predictive coding lets the brain predict upcoming musical events and reduce prediction errors. These processes start in the brainstem and extend through the auditory cortex, supporting music perception, auditory-motor coordination, and response to musical surprises.

What Happens in the AC

The auditory cortex (AC), located in the brain's temporal lobe, plays a key role in interpreting the auditory information received by the ear. After neural impulses travel through the brainstem and thalamus, they reach the AC, where they are further processed to create conscious auditory experiences. This section explores the structure of the auditory cortex, how sound is perceived and interpreted in various cortical areas, and the neural mechanisms that support our musical perception, memory, and imagination.

AC Regions

The auditory cortex is organized into approximate regions that process different aspects of sound. It is divided into three areas: the primary auditory cortex (A1), the belt region, and the parabelt region. These regions work together to process sound, with each part specializing in different functions. The primary auditory cortex (A1) is the central area of the auditory cortex and is most directly involved in the initial analysis of sound. Surrounding A1, the belt and parabelt (more in the lateral direction) regions contribute to more complex processing, including the interpretation of more abstract sound features and memory, such as meaning and context.

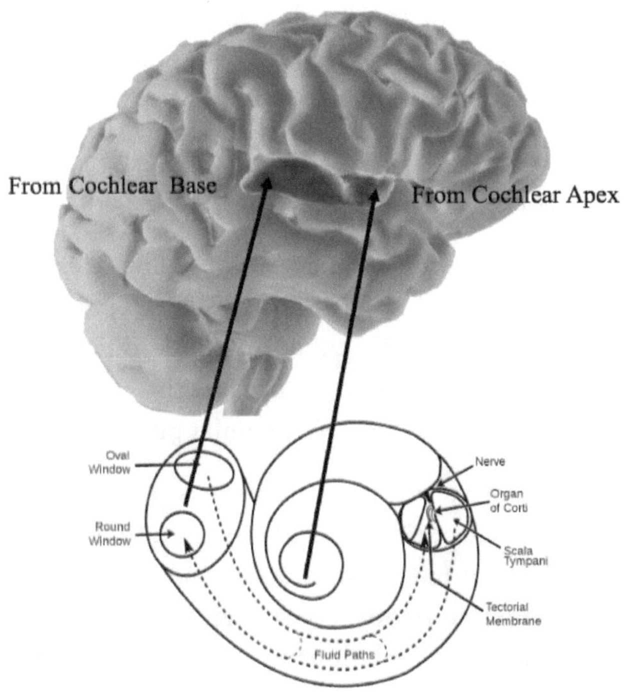

Figure 6.5. The tonotopic map, which, for the most part, preserves the frequency-to-position relationship from the cochlea, plays a crucial role in processing auditory information. The brain outline is based on the 3D atlas from Neurotorium, https:// neurotorium.org/tool/brain-atlas/.

Auditory Frequency Mapping

As previously mentioned, the auditory system's key feature is its ability to detect sound frequencies, which enables us to perceive pitch. The previous chapter shows that the cochlea acts as a spectrum analyzer, breaking down incoming sound into its component frequencies, which are then sent to the auditory cortex.

The organization of the cochlea is mainly preserved in the primary auditory cortex (A1). Different sound frequencies activate specific areas of A1, forming a spatial map of sound frequencies. Low-frequency sounds are represented in one region of A1, while high-frequency sounds are mapped in another. This tonotopic map

(Figure 6.5) reflects how the cochlea encodes the pitch of sounds. However, some mappings reported in the literature may differ from Figure 6.5, depending on how accurately the exact locations of the auditory cortex are identified.

Pitch Sensitivity in the AC Core

The core area of the auditory cortex, which includes A1, is highly sensitive to pitch and plays a key role in processing musical information. This region is especially tuned to subtle differences between frequencies, allowing us to perceive not just the pitch of a sound but also the fine details that differentiate similar pitches.

Furthermore, statistical learning influences how we perceive and comprehend pitch. Over time, our brains learn to recognize musical patterns in auditory input, such as the harmonic relationships often accompanying musical sounds. Consequently, we develop a nuanced understanding of pitch relationships and their interactions in musical contexts. This process of statistical learning improves our ability to identify familiar melodies and tunes.

However, one intriguing aspect of auditory processing is that the brain may not always be able to discern octaves perfectly. The missing fundamental phenomenon highlights the brain's tendency to group harmonics and perceive a single fundamental pitch, even if that fundamental frequency is absent. This phenomenon demonstrates the full operation of SLPC's theory and the control system modeling, as introduced in Chapter 4.

Beginning of Perception

Once the auditory information is processed in the auditory cortex, it undergoes a transformation that allows the brain to interpret the sound or music in a neural language different from simple neural impulses. After this transformation, the incoming auditory neural impulses, which consist of frequency, amplitude, and temporal patterns, are no longer represented in their raw form. This process involves recognizing familiar sounds, such as voices, music, and environmental noises, and linking them to past experiences or contexts. Ultimately, perception creates a higher-level representation of the sound, enabling the brain to understand the timing, pitch, melody, and harmony of the music. The neural impulses that once mirrored the incoming sound are now converted into an abstract representation, which is then combined with other sensory inputs and cognitive processes to form perception. As this processed information travels through the brain, it establishes the foundation for our conscious perception of sound, whether it is music, speech, or any other auditory experience.

The auditory cortex is shaped like an elongated football, as shown in Figures 6.2 and 6.3, and is surrounded by outer layers called the belt and parabelt. It tilts downward from the back of the brain, with the upper part being dorsal and the lower part ventral. After processing musical information, the AC sends the results upstream to higher areas for further analysis and interpretation. Diffusion MRI technology has revealed neuron bundles beyond the auditory cortex, guiding the flow of information from the dorsal and ventral areas in specific directions. The next chapter will discuss where these pathways end. This chapter will explain how the AC analyzes incoming information in the dorsal and ventral areas and prepares it to flow further along the pathway.

As musical information moves farther from the core of AC (A1), the auditory cortex starts to process more complex sound features separately, such as rhythm and spectral data. The dorsal part of the auditory cortex becomes increasingly responsive to the temporal aspects of sound, including rhythm, timing, and modulation. This region is especially important for processing the timing of musical notes and speech syllables. It helps us perceive the tempo of music, the timing of speech sounds, and other time-dependent characteristics of auditory stimuli.

Conversely, the ventral area of the auditory cortex is more sensitive to the spectral features of sound, including its audio frequency contents and spectral shapes. This region is crucial for processing the overall spectral qualities of sound, enabling us to distinguish between various types of sounds, including musical instruments, human voices, and environmental noises.

The balance between the dorsal and ventral regions of the auditory cortex enables us to perceive both the rhythm and melody of music, allowing our brains to comprehend musical structures. This dual processing is essential for understanding music and speech, as rhythm and pitch are vital for accurate perception, comprehension, and cognition.

Ventral Area Tasks

This area is involved in processing musical information sent to the auditory ventral stream (AVS). The ventral stream, responsible for "what" processing, focuses on identifying and recognizing auditory objects. Here are the specific tasks performed by the auditory cortex in response to music, directed toward the auditory ventral stream.

Pitch Processing: Processing the frequency components of sounds is vital for recognizing musical pitch and melody. It helps identify and represent pitch relationships, which are then sent to the ventral stream to support the recognition of musical intervals, scales, and melodies.

Timbre Recognition: The auditory cortex encodes the timbre of sounds, which refers to the quality, or color, of a musical sound that allows differentiation between different instruments or voices playing the same note.

Harmonic Structure and Chord Recognition: The auditory cortex processes the harmonic structure of music, identifying how different notes or chords relate to one another. This includes recognizing harmonic progressions, consonance, and dissonance.

The auditory cortex processes melodic and tonal sequences, including the perception of musical phrases or motifs. It also helps recognize a musical piece's overall tonality (e.g., major or minor scale).

Musical Context and Semantic Processing: The auditory cortex also processes musical context and semantic information, such as the emotional tone of music (e.g., happy, sad) and the perceived intention behind the musical performance.

Once the auditory cortex processes basic features like pitch, timbre, and rhythm, these are passed along to the ventral stream, especially to regions such as the anterior (front) superior temporal gyrus (aSTG), the temporal pole, and the inferior frontal gyrus, which are involved in higher-level processing and object recognition.

<u>Dorsal Area Tasks</u>

The auditory cortex is essential for processing music directed toward the auditory dorsal stream (ADS). The dorsal stream is often called the "where" or "how" pathway and plays a role in the spatial processing of sound, motor-related actions, and real-time sensory-motor integration. This pathway enables us to determine where sounds originate from and how to respond to them in a motor context. In music, the auditory dorsal stream is essential for locating sound sources, coordinating motor actions with music (such as playing instruments or dancing), and combining auditory cues with other sensory inputs. Here's a breakdown of the tasks the auditory cortex performs related to the auditory dorsal stream in the context of music.

Sound Localization: The auditory cortex processes spatial cues such as interaural time differences (ITDs) and interaural level differences (ILDs), which help localize sounds in space. In music, this allows listeners to determine the direction of different sound sources, such as distinguishing between instruments in an orchestra or the location of a sound in a live music performance.

Rhythm and Temporal Pattern Processing: This area of auditory cortex processes rhythm, timing, and temporal patterns in music. It detects regularities in musical time, such as beats, tempo, and syncopation. These features are essential for motor coordination, like tapping along with the beat or dancing to music.

Auditory-Motor Integration: The auditory cortex processes auditory feedback from music performances or movements. This includes input from instruments, vocal sounds, and rhythm, which must be coordinated with motor actions (e.g., playing a guitar or drumming).

Action Planning and Timing: This part of the auditory cortex helps process "temporal cues" in music, such as rhythmic patterns or musical phrasing. These cues are essential for anticipating and planning movements in musical performance.

The auditory cortex processes auditory feedback during music performance, which is essential for adjusting motor actions in real time. Musicians constantly monitor their playing and adapt based on auditory feedback (e.g., adjusting pitch, rhythm, or dynamics).

Ensemble Performance Coordination: When musicians perform together, the auditory cortex processes each individual's sounds and the overall output. This involves complex integration of auditory signals from different sources, such as various instruments in an ensemble.

After the auditory cortex processes basic features like rhythm and timing, this information is sent to the posterior (rear) regions, such as the posterior superior temporal gyrus (pSTG) and related parietal areas. These regions are crucial for spatial and motor processing. They then relay information to motor areas, such as the premotor cortex, for motor control and execution.

SLPC in the AC

Statistical Learning in AC

More advanced statistical learning processes are involved as sound information moves through the brainstem and reaches the auditory cortex. The auditory cortex learns the temporal and spectral features of sounds and the sequential patterns typical of music, such as harmonic progressions, melody patterns, tempo, and rhythm.

Neurons in the primary auditory cortex (A1) and other cortical areas, such as the superior temporal gyrus (STG), become attuned to recurring music features—such as detecting repeated musical phrases, beats, or patterns of consonance and dissonance.

This learning is probabilistic, meaning the auditory system constantly refines predictions about what musical events will occur next, enabling it to process and integrate future sounds more effectively.

Predictive Coding in AC

In the auditory cortex, predictive coding operates at a higher level of complexity. The brain predicts upcoming musical events, such as note sequences, melodic contours, or harmonic transitions.

For example, based on previous musical experiences (or statistical learning), the auditory cortex predicts musical progression, such as anticipating a specific chord change or rhythm pattern. When these predictions match incoming sounds, processing becomes more efficient. However, when there is a prediction error, like an unexpected shift in rhythm or harmony, the auditory cortex updates its prediction model to accommodate the new pattern.

Predictive coding also helps motor responses to music, like syncing movements with a beat. The dorsal auditory stream, which processes sound for motor coordination, uses predictive coding to match movement with musical timing (e.g., knowing when to tap your foot with a rhythmic beat).

Musical Imagery and Earworms

As our understanding of neurological musical processing deepens, it is intriguing to observe that it can start without external music stimuli. The auditory cortex can "hear" the music through our internal imagery of it.

This experience is triggered by the often-invoked feedforward processes from the higher levels of music processing, which send an error signal to the lower levels, suggesting that different stimuli are present. This phenomenon is often experienced as "earworms," where a catchy melody or tune becomes stuck in one's mind and repeats involuntarily. Interestingly, the brain's response to musical imagery resembles its reaction to actual music, even when no stimulus is present. Researchers can detect musical neural activity in the auditory cortex using sensitive equipment, even when music is only imagined. If someone enjoys listening to the same music, they can experience similar enjoyment by imagining it.

This phenomenon of feedforward processing shows how high-level cognitive processes, including memory, emotion, and movement, help shape musical perception and understanding. Imagining music activates the same neural pathways that are engaged when listening to real music, indicating that the auditory cortex is involved in a wide range of auditory experiences, both internal and external.

Music and Emotion in Music Imagery

Music's imagery is closely tied to memory and emotion, which explains why certain songs or melodies trigger strong recollections or feelings. The brain's capacity to recall a song based on a specific emotional or situational context demonstrates the deep link between memory and auditory perception. This relationship emphasizes the complex connection between sensory processing, memory systems, and emotional reactions in the brain.

Music Preprocessing Control System Modeling

The earlier sections describe how the auditory pathways and cortex respond to musical stimuli. Strong scientific evidence supports the idea that these two parts work in tandem with feedback/feedforward loops. As discussed in Chapter 4, the SLPC process can be represented as an engineering control system for clarity and consistency. Figure 6.6 shows a nested control system for the auditory pathways and cortex, with parts, control functions, feedback, and feedforward paths explained in neuroscientific terms. Also, the connections are arranged in loops within other loops, which is why they are called "nested" loops.

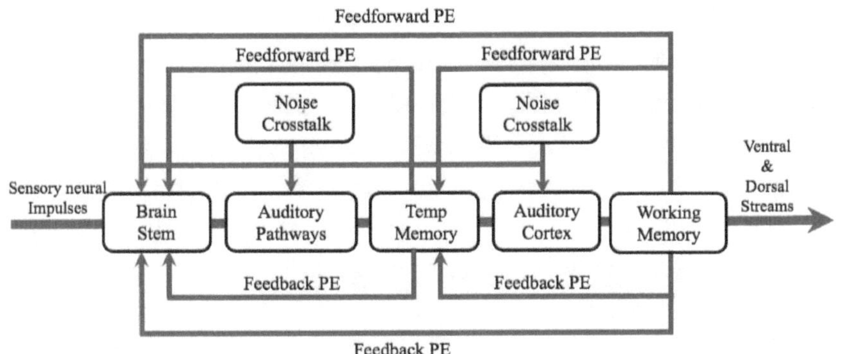

Figure 6.6. This figure describes a nested control system for the auditory pathways and cortex, where the parts, control functions, feedback, and feedforward paths are defined in neuroscientific terms. The thicker lines through the middle are the primary signal lines, whereas the thinner lines are the feedback and feedforward loops.

The neural signals on the left side of the figure serve as input to this neural control system. The far right is the output, which contains processed signals such as spectral information, timing, rhythm, harmonies, melodies, and other relevant details. The systems involved in this information processing chain include the auditory pathways and the auditory cortex.

The first part appears clear, but research shows that the neural connections to the auditory cortex are bidirectional: they include upward connections toward higher brain levels and downward signaling toward lower levels of the brain hierarchy. These up-and-down connections help transmit neural impulses nearly intact all the way to the AC, along with signals that carry predictive error information. The upstream flow allows the AC to interpret information based on raw data in an undisturbed and timely manner. The downstream flow transmits information from comparing incoming patterns with the previously stored music image. (It might not be the exact music piece, but it could be the patterns we learn through tradition.) This feedback loop inevitably causes a delay due to the time needed for the signals to be compared with memory and sent to lower neural levels.

As we listen to music, the memory detects differences between incoming sounds and stored models, instructing the brainstem mechanism to adjust what we hear. Conversely, when the memory recognizes a familiar musical pattern and primes the response based on experience, no predictive error occurs. At this point, the neural state remains in its lowest energy configuration. When a difference is detected, the predictive error suppresses it and sends it downstream to the brainstem before the expected music occurs. This process alters hearing, perception, and memory in the hippocampus. This loop constitutes the feedforward mechanism.

Interestingly, the brainstem, very early in the chain of the music cognition process, has been observed to distinguish between consonant and dissonant tones. This cognition, typically happening at a high hierarchical chain, exemplifies the long reach of the SLPC feedforward loop in our brains. On the other hand, rigorous physical study of consonance could be built into the neural impulses, as will be addressed in Chapter 9.

The second system in Figure 6.6 describes the AC operation, which processes the incoming raw data outlined in the last section. This includes pitch recognition, timbre recognition, pattern recognition, harmony and chord recognition, sound localization, temporal and rhythm identification, and auditory-motor integration.

As explained earlier, ascending and descending connections enable information flow in their respective directions. The faster lemniscal neurons primarily control the auditory pathway to the AC. The descending pathway uses non-lemniscal neurons to signal changes, such as the predictive error, and is less affected by the timing of the messages.

These two systems are interconnected, as they are linked through similar structures that exchange information in both directions. Therefore, the combined control system model offers a more complete description of the music stimuli before they are assigned meaning, which is a cognitive process discussed in the next chapter.

It is important to reiterate that the division of this aspect of auditory processing is more traditional than logical. To clarify, the control system model can be seen as a continuum with infinitesimal feedback and feedforward pathways that help interpret the messages in the music.

The upcoming chapters will cover the perception and control system models similarly and will not go into as much detail as this chapter. However, the concept of nested control systems will remain the same.

Summary

This chapter discusses how musical information moves from the ears to the auditory cortex. The first part describes the pathway to the auditory cortex, and the second part focuses on AC's analytical functions. Keep in mind that this division is artificial and may be adjusted as more detailed connections are understood. Future research can shed light on these topics.

The auditory pathway primarily transmits information from the cochlea to the auditory cortex. Along the way, it also plays a role in statistical learning and predictive coding. Notably, near the lower end of the brain hierarchy, the brainstem can integrate predictive errors when necessary to shape the perception of sounds based on stored consistent experiences in the memory system. Description of feedback and feedforward control system modeling can be beneficial from the start of auditory training. Looking further ahead, such feedback and feedforward models can extend to include group musical activities, as will be discussed in Chapter 8.

The auditory cortex is a highly specialized and organized brain region essential for analyzing and interpreting sound. From frequency-to-position mapping in the cochlea to the complex processing of pitch, rhythm, and timbre in the auditory cortex, this network of regions allows us to experience and understand the auditory world.

Whether processing music, speech, or environmental sounds, the auditory cortex converts raw sensory information into meaningful experiences. Through statistical learning, memory, and emotion, the AC helps us perceive and recognize music in ways that are deeply tied to our prior experiences and expectations.

In sum, the combination of pathways and AC serves two functions. First, it interprets the received information and processes it to its respective connections. Second, the combination operates in the fashion of the SLPC, enabling us to enjoy music that is familiar, new, and surprising.

Naturally, this part of the neural process's input is the auditory neural impulses. The output is the decoded information needed for subsequent activity, while minimizing the energy required to maintain it, as dictated by the information flow and its interaction with the next neural process.

Once auditory information is processed and decoded in the auditory cortex, it is sent to other parts of the brain for further processing and integration. The two main pathways are the ventral stream (the more "what" pathway) and the dorsal stream (the more "how" pathway).

Both pathways connect to brain regions involved in emotion and motor control, allowing us to respond emotionally by feeling triggered by melodies and to move physically by following the rhythms. The next two chapters focus on these topics.

Bibliogrpahy

1. "On the action potential as a propagating density pulse and the role of anesthetics," Heimburg, T., Jackson, A.D. *Biophys. Rev. Lett.* **2**: 57–78 (2007).
2. "The auditory cortex," Bendor, D., & Wang, X. *Current Opinion in Neurobiology*, **15**(4), 500-507 (2005). https://doi.org/10.1016/j.conb.2005.07.008
3. "Information flow in the auditory cortical network," Hackett, T. A. *Hearing Research*, **271**(1-2), 133-146 (2011). https://doi.org/10.1016/j.heares.2010.07.001
4. "The cortical organization of speech processing," Hickok, G., & Poeppel, D. *Nature Reviews Neuroscience*, **8**(5), 393-402 (2007). https://doi.org/10.1038/nrn2113

5. "The dorsal and ventral auditory pathways: Neuroanatomical and neuropsychological evidence," Jäncke, L., et al. *Cortex*, **47**(3), 323-327 (2011). https://doi.org/10.1016/j.cortex.2010.01.014

6. "The role of the superior olivary complex in the auditory processing of sound localization," Kavanagh, G. L., & Kelly, J. B. *Science*, **212**(4490), 1154-1156 (1981). https://doi.org/10.1126/science.7190491

7. "A duplex theory of pitch perception," Licklider, J. C. R. *Experientia*, **7**(12), 128-133 (1951). https://doi.org/10.1007/BF02150799

8. "Statistical learning of tone sequences by human infants and adults. *Cognition*," Saffran, J. R., et al. **70**(1), 27-52. (1999)., https://doi.org/10.1016/S0010-0277(98)00075-4

9. "When the brain plays music: Auditory-motor interactions in music perception and production," Zatorre, R. J., et al. *Trends in Cognitive Sciences*, **11**(5), 252-258 (2007). https://doi.org/10.1016/j.tics.2007.03.004

10. "Spectral and Temporal Processing in Human Auditory Cortex," Zatorre, R. J., & Belin, P. *Cerebral Cortex*, **11**(10), 946-953 (2001).

11. "Brain Organization for Music Processing," Peretz, I., & Zatorre, R. J. *Annual Review of Psychology*, **56**, 89-114 (2005).

12. "Functional Specialization of the Human Auditory Cortex," Wessinger, C. M., & Buxton, R. *Cognitive Science*, **25**(3), 357-373 (2001).

13. "Lateralized Processing of Pitch in the Human Auditory Cortex," Tervaniemi, M., & Hugdahl, K. *Brain Research Reviews*, **43**(2), 256-264. (2003).

14. "Music and Emotion: From Theories to Brain Imaging," Brown, S., Martinez, M. J., & Parsons, L. M. *Psychology of Music*, **34**(4), 413-440 (2006).

15. "Two Functional Components of Perception and Production

of Musical Timbre," Halpern, A. R., & Zatorre, R. J. *Journal of Cognitive Neuroscience*, **11**(3), 292-303. (1999).

16. "Towards a Neural Basis of Music Processing." Koelsch, S. *Frontiers in Psychology*, **2**, 110 (2011).

17. "The Perception and Cortical Representation of Speech," Griffiths, T. D., & Warren, J. D. *Trends in Cognitive Sciences*, **6**, 320-327 (2002).

18. "Music, language, and the brain." Patel, A. D. *Oxford University Press.* (2008).

19. "The unique role of the non-lemniscal pathway on stimulus-specific adaptation (SSA) in the auditory system," G. V. Carbaja and M. S. Malmierca, Adaptive Processes in Hearing, *Proceedings of the International Symposium on Auditory and Audiological Research at Nyborg, Denmark*, **6** (2017).

Chapter 7

Music Perception

In the previous chapter, we examined how the auditory cortex takes in raw acoustic data and converts it into neural signals with higher-level meanings. These initial steps, similar to pre-processing in a computer system to ready the data for further analysis, represent the third phase in the "phased-naturalism" framework introduced earlier (see Figure 1.1). Now, we proceed to the next phase: full music perception.

This stage no longer deals with raw acoustic data. Instead, it uses pre-processed information to form messages in the brain's internal language, which is understandable to other higher-level brain regions. Music is no longer just sound; it becomes organized information with meaning and emotional nuance.

Before we proceed further into this stage of music cognition, it is important to recognize that we have reached a different level in the brain's processing hierarchy. This stage also marks the boundary between phases, as it involves advanced brain organization and functions unique to humans. Specifically, the neural processes mainly occur in the cortices, which are thinner and simpler in mammals other than humans.

In the previous chapters, we started at the most basic level, where airwaves are transformed into electrical signals by the ear and sent through the auditory pathways to the auditory cortex. This initial processing keeps a map of sound frequencies, called tonotopy, as shown in Figure 6.2. Up to that point, the brain is still handling the physical aspects of sound.

But once the auditory cortex sends signals to higher brain regions, the information undergoes a profound transformation. It's no longer just about frequencies and sound waves—it becomes abstract, symbolic, and emotionally meaningful. The signals are now encoded in ways that are fundamentally different from the raw input that entered through the ears.

In humans, this next step involves some of the most advanced parts of the brain, particularly the prefrontal cortex, the most recently evolved region. This area is responsible for our highest cognitive functions: abstract thinking, reasoning, language, and complex problem-solving. The expansion of gray matter, essentially a denser network of neural processing, gives humans a unique capacity for reflection, imagination, and cultural creation. It's what enables us not just to perceive music, but to shape it, remember it, and be moved by it.

From this perspective, earlier brain regions deal with the basics, acoustic signaling, and tone detection. In a simpler brain like other mammals, this might be the end of the road. But in humans, music continues its journey upward, entering a new phase of interpretation. We don't just hear music, we experience its structure, patterns, and emotional resonance.

In this chapter, we follow the pathways by which auditory information flows through the brain's architecture, the ventral and dorsal streams. Along the way, we'll explore how the brain identifies musical patterns, generates emotional responses, and links music to memory, movement, reward, and meaning.

We'll also revisit SLPC (Statistical Learning and Predictive Coding) and compare it with a control system framework to show how the brain constantly matches new musical input with past experiences. These mechanisms help set the stage to explain how we recognize melodies, feel rhythm, and even get chills when listening to powerful musical moments.

The Auditory Ventral and Dorsal Streams

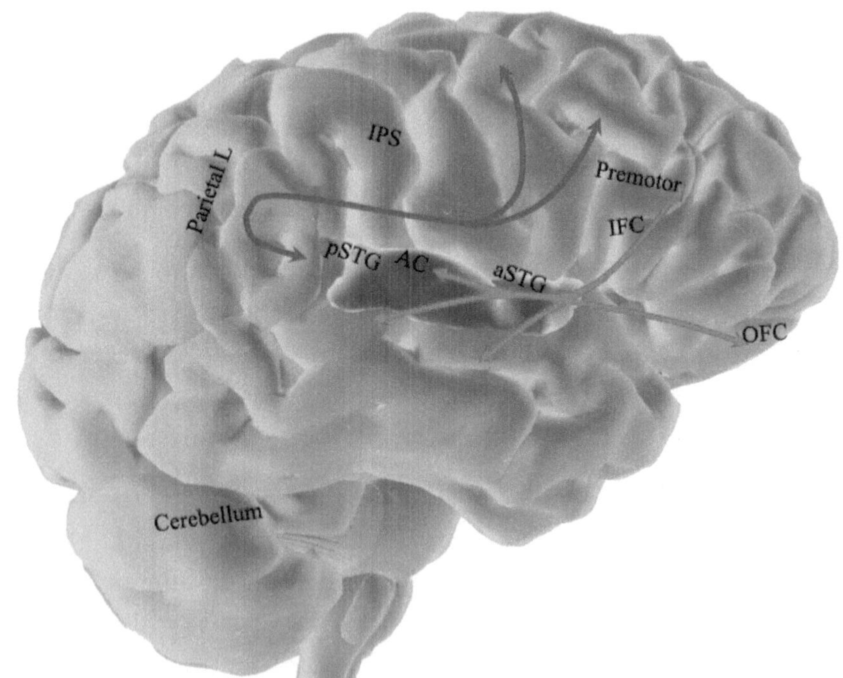

Figure 7.1. *This figure offers a simplified side view of the right hemisphere of the human brain, highlighting the key regions involved in music processing. It focuses on the two major neural highways that carry auditory information beyond the auditory cortex: the ventral and dorsal streams. The brain outline is based on the 3D atlas from Neurotorium, https://neurotorium.org/tool/brain-atlas/.*

Figure 7.1 presents a simplified side view of the right hemisphere of the human brain, highlighting the main regions involved in music processing. It emphasizes the two major neural pathways that transmit auditory information beyond the auditory cortex: the ventral and dorsal streams. Typically, the ventral stream connects the temporal lobe to the frontal lobe, while the dorsal stream links the AC to the parietal lobe, premotor cortex, and inferior frontal lobe. The auditory cortex (AC), situated in the upper part of the temporal lobe just below the Sylvian fissure and above the superior temporal gyrus (STG), serves as the launching pad for both pathways to the reward system and the pre-motor cortex.

The lines shown in the figure are derived from diffusion tensor imaging (DTI), a type of MRI that tracks water molecules' movement along nerve fibers. This technique enables researchers to map the brain's structural connections. However, the full picture is much more complex than the figure indicates because the pathways also pass through the internal regions of the brain. For clarity, we'll focus here on the main pathways involved in processing musical information as it moves up the neural hierarchy.

It's important to remember that the human brain is a three-dimensional structure. The side-view diagram is a simplification; it doesn't show the midbrain's lateral (side-to-side) neural connections, especially those linking to the brain's reward centers, which are located deep in the midbrain. We'll examine those connections more closely in the next chapter.

The Auditory Ventral Stream (The Why Pathway)

The auditory ventral stream begins at the AC and progresses forward through the anterior superior temporal gyrus (aSTG). From there, it crosses the Sylvian fissure to reach the inferior frontal gyrus (IFG). This stream plays a role in identifying and interpreting sound content, especially musical tones and patterns, which is why it's often called the "why" pathway.

The ventral stream also connects to the orbitofrontal cortex (OFC), a region involved in emotion and decision-making, as well as to the amygdala, which plays a central role in emotional memory. It also reaches the lower part of the striatum, namely the ventral striatum and nucleus accumbens. These connections suggest that emotional evaluation occurs early in the music-processing pathway. Some branches also extend to the hippocampus, an important structure for forming and retrieving both short-term and long-term memories of musical experiences. In the diagram, these ventral stream connections are shown in green.

The Auditory Dorsal Stream (The How Pathway)

The auditory dorsal stream begins at the back of the auditory cortex, specifically in the posterior superior temporal gyrus (pSTG). It extends upward through the parietal lobe, connecting with the intraparietal sulcus (IPS) and continuing toward the premotor cortex. Another branch also reaches the inferior frontal cortex (IFC) and finally the dorsal striatum, which is part of the brain's reward circuitry.

This stream primarily handles sound localization and spatial awareness. Since it assists in determining where sounds come from and how we engage with them, it is often referred to as the "how" pathway.

In the human brain, especially in the left hemisphere, the auditory dorsal stream also plays a role in speech production, repetition, lip-reading, and phonological memory. On the right hemisphere, it mainly processes musical features such as melodic contour, beat, and rhythm, especially those not related to language. It helps us understand music through movement, space, and timing. In the figure, dorsal stream connections are shown in red.

The Auditory Ventral Stream

At the beginning of musical perception, the auditory ventral stream performs several basic functions. As auditory information moves up the brain's hierarchy, it is processed with increasing complexity. The auditory core analyzes tones and pitches; the parabelt processes melodic sequences; and the auditory cortex (AC) sorts through raw auditory inputs.

The ventral stream's role in music involves decoding musical features like scales, melodies, and, to a lesser extent, rhythms. Specifically, the anterior superior temporal gyrus (aSTG) interprets short-term pitch patterns, such as melodies, before sending that information to the inferior frontal cortex (IFC). It also supports our ability to encode and retrieve musical experiences, aiding mnemonic processing, which helps our brain remember musical patterns.

Specialized AVS Functions

The auditory ventral stream specializes in recognizing, categorizing, and interpreting musical elements related to sound quality and structure.

Melody Recognition: It aids us in identifying familiar tunes by recognizing pitch sequences and relationships, whether in instrumental or vocal music.

Harmony and Chord Recognition: It helps the brain identify and interpret harmonic structures, such as chord progressions and their emotional impact, distinguishing between consonance and dissonance.

Rhythm and Temporal Structure: While the dorsal stream mainly processes rhythm, the ventral stream helps by identifying rhythmic patterns, tempo shifts, and syncopation in a musical piece.

Timbre Identification: Timbre is the "color" of a sound that helps us distinguish one instrument or voice from another. The ventral stream enables us to tell apart a violin and a flute, even when they play the same note.

Musical Emotion and Expressiveness: The ventral stream detects emotional tone in music, such as tension, relaxation, joy, or sadness, by analyzing dynamics and intensity.

Genre and Style Recognition: It detects patterns typical of different genres (jazz, classical, rock) by analyzing features such as melody, rhythm, harmony, and timbre.

Instrumental Sound Recognition: This stream helps us identify which instrument is playing a particular note or passage, whether it's a piano, trumpet, or drum.

Memory and Familiarity: It links to memory centers to help us recognize songs we've heard before and detect changes in familiar tunes.

Musical Structure and Form: The ventral stream helps us understand how a musical piece is organized, such as recognizing verse-chorus structures in pop music or movements in a symphony.

Voice and Lyrics Recognition: For vocal music, the ventral stream helps identify lyrics and singers' voices by integrating phonetic cues with melody and rhythm.

These capabilities rely on interactions between the auditory cortex and other regions of the brain, especially the aSTG, IFC, and nearby areas in the temporal lobe. Together, they decode the layered music experience.

Avs's Connection to Reward Systems

The auditory ventral stream not only processes musical elements; it also connects with the brain's reward circuitry, especially via the orbitofrontal cortex (OFC). This network includes areas like the nucleus accumbens and the ventral tegmental area (VTA), which are associated with pleasure, motivation, and reinforcement learning (see Figure 8.1).

The ventral stream delivers processed auditory information, such as melodic lines or harmonic shifts, to the OFC. The OFC evaluates this input based on previous experiences and emotional associations, determining how enjoyable or emotionally moving the music is and assigning a value for the reward system to interpret.

If the music is perceived as rewarding, the OFC signals the nucleus accumbens and VTA, triggering dopamine release, the brain's pleasure chemical.

This creates a positive feedback loop (positive feedback is not part of the SLCP operation): the more rewarding the music, the stronger the emotional response, which in turn reinforces our desire to listen to it again. This mechanism helps explain emotional chills, euphoric highs, or nostalgic waves triggered by specific musical phrases.

<u>AVS's Connection to Memories</u>

The ventral stream also plays a crucial role in linking music to memory systems, mainly via its links to the hippocampus, parahippocampal gyrus, and amygdala.

Music Recognition: Helps us identify familiar songs by retrieving long-term memories, including their structure, lyrics, or emotional significance.

Emotional Memory: Connects with the amygdala to associate songs with emotional experiences—explaining why a particular tune can immediately bring back a life event.

Encoding and Recall: collaborate with the hippocampus to encode new musical experiences, such as learning a song or melody.

Memory Consolidation: Supports long-term storage by integrating musical memories with other contextual details, such as your location or feelings.

Together, these functions make music not just an auditory experience but a personal and emotional one that stays with us.

<u>SLPC for AVS</u>

The ventral stream operates according to the theories of Statistical Learning and Predictive Coding (SLPC). In this hierarchical system, lower levels handle simpler, faster processes (e.g., pitch detection) and higher levels manage more complex tasks (e.g., melody recognition or emotional interpretation).

As musical information travels upward through the aSTG to the IFC, the brain makes predictions about what it expects to hear next. These predictions are constantly tested against incoming data.

Suppose a prediction error (PE) occurs, meaning the actual input differs from what was expected. In that case, non-lemniscal neurons send a signal back down the hierarchy to adjust predictions or update memory. This feedback/feedforward loop enables the brain to learn from each musical encounter and refine its future responses.

What happens when this SLPC loop breaks down for some reason? For individuals with amusia, a condition marked by difficulty recognizing music, this control loop is disrupted. MRI scans show reduced white matter connectivity between the aSTG and IFC, meaning that musical data may not flow efficiently. At the same time, increased gray matter in the IFC indicates that the brain is still trying to process the insufficient data but lacks the necessary upstream information to do so effectively. These individuals usually retain language and analytical reasoning skills, supporting the idea that music processing relies on a specialized, hierarchical system. However, if the feedforward or feedback mechanism fails, music recognition becomes impaired.

Some research suggests possible genetic roots for amusia, including mutations in the 22q11.2 region or variations in genes such as AVPR1A, SLC6A4, and loci on chromosome 4. While the genetic understanding remains incomplete, the structural evidence strongly supports the hierarchical and predictive nature of AVS processing.

This amusia case study provides further evidence that the SLPC theoretical framework is valid and can serve as a solid foundation for neuroscience research.

AVS Control System Modeling

The control system model for the auditory ventral stream is summarized in Figure 7.2. Here's how it works conceptually, based on the scientific background introduced in Chapter 4. As signals from the lower (ventral) part of the auditory cortex enter the stream, they trigger the following ascending neural signals (the input). This process continues as the ventral stream analyzes melodic and spectral content (pitch, harmony, melody) and assesses their emotional significance. This information is then routed to the IFC (for content analysis) and the OFC (for emotional evaluation).

Along the way, higher neural regions compare their input with short-term and long-term memory representations of past musical experiences. If the new input matches expectations, the system stays stable. If not, prediction errors cause the brain to update its internal models stored in memory, either through feedback or feedforward pathways to lower parts of the system, thus completing the control system loops as shown by the lighter lines.

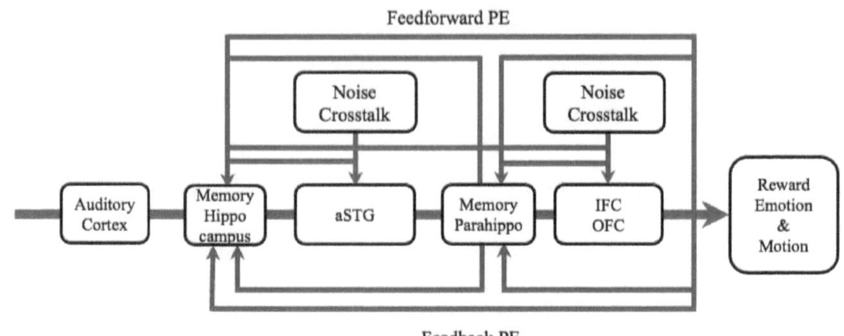

Figure 7.2. The control system model for the auditory ventral stream is summarized in this figure. The thicker lines through the middle are the primary signal lines, whereas the thinner lines are the feedback and feedforward loops.

The Auditory Dorsal Stream

The auditory dorsal stream shares some organizational features with the dorsal stream in the visual system. However, instead of creating spatial maps of visual objects, the auditory dorsal stream constructs a temporal and spatial model of sound. It processes the arrangement of musical events over time—such as rhythm, beat, and pitch movement—while also linking sound to motor actions.

Working closely with the auditory ventral stream, the dorsal stream tracks the timing, sequence, and flow of musical elements and connects them with physical movement, memory, and spatial awareness. This connection is essential for understanding rhythm, performing music, and collaborating with other musicians.

The dorsal stream's interaction with theta brainwaves is especially important. These waves help encode and retrieve musical memories, often by replaying past experiences or mimicking rhythmic variations. This gives the ADS flexibility in how we interpret time, tempo, and rhythmic flow. The theta waves' impact will be explored further later.

Specialized ADS Functions

What follows is an itemized list of functions the auditory dorsal stream specifically performs for music.

Spatial localization of sound sources: The auditory dorsal stream is essential for identifying where sounds originate in space, which is critical for understanding the source of musical sounds. This is especially important during performances with multiple instruments, as it helps listeners determine the placement of each instrument within the sound field (e.g., a piano on the left, a guitar on the right).

Movement and Gesture Interpretation in Music: The dorsal stream aids in perceiving and interpreting musical movements, such as a performer's physical gestures (e.g., a conductor's gestures or a violinist's bowing technique). It integrates auditory information with motor processing, helping the brain connect sound changes to physical actions.

Tempo Perception and Timing: The dorsal stream is vital for recognizing the timing and rhythm of music, especially during dynamic or live performances. It aids in tracking beats, tempo shifts, and syncopation, which are crucial for staying in sync with the music and coordinating movement during musical activities (e.g., dancing or playing an instrument).

Pitch Tracking and Tonal Movement: The dorsal stream processes variations in pitch and tonal movement, which are essential in music for understanding melody direction, key changes, and harmonic shifts. It monitors pitch contours and guarantees accurate perception of melodic movement (upward or downward) and the intervals between notes.

Audio-Motor Coordination: The auditory dorsal stream helps coordinate motor actions with musical sounds. For example, it assists musicians in playing in sync with others by integrating auditory input with the motor control needed to play an instrument. Likewise, it plays a role in aligning physical movement with rhythm during activities like dancing or conducting.

Sound Movement and Dynamic Processing: It analyzes how we perceive sound movement, which can occur in real-time (such as stereo or surround sound panning) or through changes in sound over time. For example, when a sound source moves around in a musical environment, the dorsal stream helps track and understand these movements.

Attention to Auditory Details in a Musical Context: The dorsal stream helps in directing attention to relevant sounds in music, like following a soloist in an orchestra or focusing on a specific instrument in a band. This attention-guiding role assists listeners in staying focused on particular auditory features that may be important for performance or enjoyment.

Auditory Feedback for Musical Performance: The dorsal stream processes auditory feedback during music performance, helping musicians adjust their playing in real time based on what they hear. This feedback loop (not the microscopic feedback) ensures that musicians can correct pitch, rhythm, and timing, maintaining coordination in group settings or with accompaniment.

Motor Learning in Musical Contexts: The dorsal stream is involved in learning motor skills related to playing an instrument or singing. It helps musicians refine their motor actions, such as finger placement on a guitar or hand positioning on a piano, based on the auditory feedback they receive during practice.

Perception of Movement in Musical Sound (e.g., Glissando or Slide): In some musical contexts, such as a glissando or pitch slide, the dorsal stream helps in perceiving the continuous movement of pitch. This ability to perceive gradual pitch changes (like sliding between notes) is especially important in genres like jazz or certain orchestral techniques.

The auditory dorsal stream works alongside the ventral stream to provide a complete auditory experience, ensuring that both recognition and action-based aspects of music are well-integrated.

ADS's Connection to the Reward System

We noted before that the auditory dorsal stream connects to the brain's reward systems via its links to the IFC and dorsal striatum. The following provides more details.

Linking the Reward System With Auditory-Motor Coordination

It is essential for integrating auditory information with motor actions, such as syncing movements to music or playing an instrument. This coordination depends on feedback and feedforward from the brain's reward systems (Chapter 8), especially the dopaminergic pathways, which are involved in motivation, reinforcement learning, and pleasure.

When a person successfully synchronizes their movements to music or plays an instrument, the brain often rewards this achievement by releasing dopamine, signaling a pleasurable outcome. The auditory dorsal stream helps predict and refine movements based on auditory cues, and when actions are well-coordinated, the brain reinforces these successful behaviors with dopamine, increasing the likelihood that the person will repeat the behavior.

Rhythmic Synchronization and Dopaminergic Reward: Research shows that syncing with music rhythm activates brain regions involved in reward, such as the dorsal striatum and ventral striatum, which are parts of the brain's reward system. These regions process dopamine, encouraging ongoing engagement with the music. As the auditory dorsal stream interprets rhythmic cues and helps coordinate motor responses (e.g., tapping along to a beat), the reward systems enhance motivation and pleasure connected to rhythmic activity.

The link between the auditory dorsal stream and reward systems is also crucial for emotional reactions to music. Music often elicits strong feelings, from joy and happiness to excitement and nostalgia. When listening to music that feels rewarding, such as pleasurable melodies or emotional pieces, the auditory dorsal stream processes the sound information, while the reward systems—especially those involving dopamine—respond to the enjoyable or engaging aspects of the music. The brain's reward pathways then produce emotional responses through dopamine release, creating feelings of pleasure and encouraging people to listen to or engage with the music. This process can increase motivation, making individuals more likely to pursue rewarding musical experiences again in the future.

Anticipation of Reward and Predictive Coding: The auditory dorsal stream's role in predictive coding also helps in expecting the emotional or rewarding qualities of music. When people listen to familiar tunes, the brain predicts the emotional reward associated with certain melodies or harmonies, activating reward-related areas in anticipation of the pleasure they expect. This anticipation is rewarding on its own, boosting feelings of engagement and emotional connection with the music.

<u>Learning and Motivation Through Music</u>

The auditory dorsal stream's connection to reward systems also plays a crucial role in motivation and learning within musical contexts. As individuals acquire new musical skills or practice rhythmic synchronization, the auditory dorsal stream helps process auditory feedback and fine-tune motor actions. When they succeed or show improvement, the reward system reinforces these behaviors through dopamine release, encouraging further learning and practice.

The auditory dorsal stream's role in auditory-motor integration means that successful coordination of music and movement will trigger dopamine release, reinforcing those motor behaviors. This loop of auditory feedback, motor adjustment, and rewarding experiences fuels the motivation to keep practicing and enhancing musical skills.

<u>Reward Systems in Social Music Interaction</u>

The connection between the auditory dorsal stream and reward systems is particularly strong during social music interactions. For example, group music-making, such as playing in a band or singing in a choir, often triggers the reward system because of the shared emotional and social experiences. The auditory dorsal streams also help process and synchronize the auditory signals within the group, while the reward system boosts social and emotional bonding, which is further strengthened by dopamine release.

Making music with others promotes social bonding and collective achievement, driven by the reward system. The auditory dorsal stream helps people stay synchronized, and the reward system releases dopamine to reinforce this successful social connection, fostering positive feelings of togetherness and shared success.

In summary, the auditory dorsal stream connects to the brain's reward systems in multiple ways, particularly through auditory-motor coordination, emotional processing, and reinforcement learning. By helping people synchronize their movements with music, predict auditory stimuli, and process emotional rewards, the auditory dorsal stream plays an essential role in music-related motivation, engagement, and learning. These links emphasize the strong connection between sensory perception, motor actions, and emotional rewards in the context of music, highlighting the powerful interaction between music and the brain's reward circuitry.

<u>How Does Theta Wave Help Music Comprehension?</u>

Theta waves are a type of brainwave with a frequency generally between 4 and 8 Hz, linked to various cognitive functions like attention, memory, and emotional processing. They are detected through

EEG by measuring electromagnetic (EM) variations over time. Theta waves are intriguing because they oscillate at the same natural EM waves known as Schumann Resonance. It is believed that human brainwaves evolved in an environment where Schumann resonance was present everywhere on Earth for at least a billion years. Typically, human theta waves are observed in brain regions related to memory and learning, such as the hippocampus and auditory processing areas.

In the context of music comprehension, theta waves originating from the auditory dorsal stream play a key role in how we process and understand music. Although the ADS is usually linked to spatial and motor functions, recent studies also indicate that the auditory dorsal stream helps with higher-level aspects of music cognition. The following describes how theta waves generated in this pathway may support music understanding.

<u>Integration of Rhythm and Timing</u>

The auditory dorsal stream is heavily involved in processing rhythmic and timing information, which are essential components of music. Theta waves are believed to play a role in synchronizing the brain's oscillatory activity with external rhythmic cues, such as a musical beat. As we listen to music, the brain may synchronize with the tempo or rhythm of the music, and theta waves help to coordinate this synchronization. This neural synchronization improves our ability to perceive and predict musical rhythms, making it easier for us to follow along and understand the structure.

For example, when listening to complex rhythms or syncopated beats, theta waves can help the brain "lock" onto the musical timing, thereby supporting our understanding of the music's flow. This is especially important for the perception and recognition of tempo and meter, which are fundamental aspects of musical comprehension.

Memory Encoding and Retrieval of Music Patterns

Theta waves are also linked to memory, especially in the hippocampus, which helps encode new memories and retrieve them. The auditory dorsal stream processes auditory information related to the temporal aspects of sound—how sounds are organized over time. Theta waves generated in this stream might help encode musical sequences and patterns into memory.

When we listen to music, we naturally recall and expect recurring patterns, themes, and motifs. Theta waves support this process by helping us recognize musical structures, such as chord progressions, rhythmic patterns, or motifs that repeat throughout a piece. By aiding in memory encoding and retrieval, theta waves help us understand music over time, as our brain can "remember" these musical patterns and anticipate upcoming elements in the song.

Emotional Processing

Theta waves are also connected to emotional processing, which is essential for music perception. Music often evokes strong emotional responses, and the auditory dorsal stream helps us interpret the spatial and dynamic features of sound that shape emotional experiences in music. Theta activity has been shown to be involved in emotional resonance and processing in response to auditory stimuli.

As we listen to music, theta waves in the auditory dorsal stream help link the movement of sound (such as crescendos or decrescendos) with emotional responses, which is a key part of understanding the emotional content of music. The rhythm and tempo of music often carry emotional significance (for example, a slow, steady beat may evoke feelings of sadness or calmness, while an upbeat rhythm may evoke joy or excitement). Theta waves may assist the brain in processing emotional information, thereby improving the overall experience and understanding of music.

Attention and Focus

Theta waves are also associated with states of focused attention, especially when we are deeply engaged in a task. In music comprehension, theta waves may help the brain stay focused on specific musical details, such as pitch shifts, harmonic changes, or the progression of musical phrases. The auditory dorsal stream's role in attention and spatial processing supports the brain in maintaining a clear understanding of the music as it unfolds, encouraging ongoing engagement with the piece.

For example, in complex or dense music, theta oscillations can help the listener focus on the most important features—whether it's the melody, harmony, or rhythmic structure—allowing for a deeper, more nuanced understanding of the piece.

In summary, theta waves originating from the auditory dorsal stream are essential for timing and rhythm processing, memory encoding and retrieval, emotional responses, focused attention, and motor activities. These functions are key to understanding music because they help the brain integrate rhythmic, emotional, and spatial elements, improve our ability to anticipate musical patterns, and connect auditory experiences with motor and emotional reactions. The result is a more comprehensive, integrated understanding of music that engages various cognitive and emotional systems in the brain.

SLPC in Dorsal Stream

In the context of music, ADS plays a key role in statistical learning and predictive coding processes. These processes assist the brain in anticipating, adapting to, and interpreting musical sounds, rhythms, and movements.

Statistical Learning in ADS

Rhythmic Structure: Statistical learning in the dorsal stream enables listeners to detect rhythmic patterns and regularities in musical sequences. Over time, they learn typical durations between beats, recurring motifs, and time signatures, which helps them predict upcoming rhythmic events and synchronize motor actions (e.g., tapping feet or playing an instrument).

Spatial and Temporal Patterns: The dorsal stream processes how sound changes across space and time, such as the movement of sound sources (in live music or soundscapes) or the development of melodies during a performance. Through statistical learning, the brain becomes attuned to these shifts, enabling anticipation of spatial movements or temporal deviations in music.

Auditory-Motor Learning: Statistical learning also helps link auditory perception to motor actions. For musicians, for example, understanding how specific musical notes or sequences relate to hand movements or finger placements depends on recognizing statistical patterns between sounds and the corresponding motor responses.

Predictive Coding in ADS

Prediction of Musical Structure: Predictive coding helps the auditory dorsal stream form expectations about music, including upcoming notes, rhythms, and harmonic progressions. For example, when listening to a familiar piece, the brain anticipates upcoming chords or note sequences based on learned patterns and past musical experiences. If the sound matches the prediction, it is processed smoothly. If there is a mismatch (such as an unexpected chord or note), the brain updates its model to include the new information.

Timing and Movement Synchronization: In musical performance, predictive coding helps anticipate beats, downbeats, and other timing features. This is essential for musicians to stay in sync with an ensemble or rhythm, enabling motor actions (like playing an instrument) to align with the sound's timing. The auditory dorsal stream predicts when specific rhythmic events (such as beats) will occur, reducing delays between hearing and movement.

Prediction of Spatial and Motor Feedback: The predictive coding framework also aids in anticipating and adjusting for spatial shifts and motor feedback. For example, when a musician shifts positions on the instrument or when a sound moves through space, the brain uses past experiences to forecast the outcomes of these movements and adapt motor actions accordingly. This is especially crucial in ensemble music, where musicians must expect the movement of sound within the group.

<u>Integrating Statistical Learning and Predictive Coding in ADS</u>

The auditory dorsal stream combines statistical learning and predictive coding to effectively process musical information. Statistical learning aids the brain in recognizing recurring patterns in music, while predictive coding updates expectations based on incoming sensory input. These processes work together to enhance sound event prediction and enable smooth auditory-motor coordination during musical performance. The interaction between these mechanisms allows musicians and listeners to anticipate and respond to musical changes in real-time, improving their experience and performance.

In the auditory dorsal stream, statistical learning and predictive coding work together to process and understand music. Statistical learning helps the brain recognize patterns in rhythm, pitch, and spatial sound features, while predictive coding allows it to anticipate and adjust to incoming musical events. These mental processes are essential for musicians to synchronize their movements with music, predict upcoming sounds, and perform in real-time auditory-motor coordination.

<u>Control System Modeling of the Auditory Dorsal Stream</u>

Figure 7.3 depicts the control system model of the dorsal stream. Although feedback and feedforward arrows are not visible, they are implicitly included to keep the diagram simple and aid understanding. The dorsal stream creates loops that connect ascending sensory information with descending predictive models. These loops allow for real-time adjustments in motor activity and perception as music plays. Through these dynamic feedback and feedforward systems, the brain fine-tunes musical interactions, whether tapping a beat, dancing, or playing an instrument.

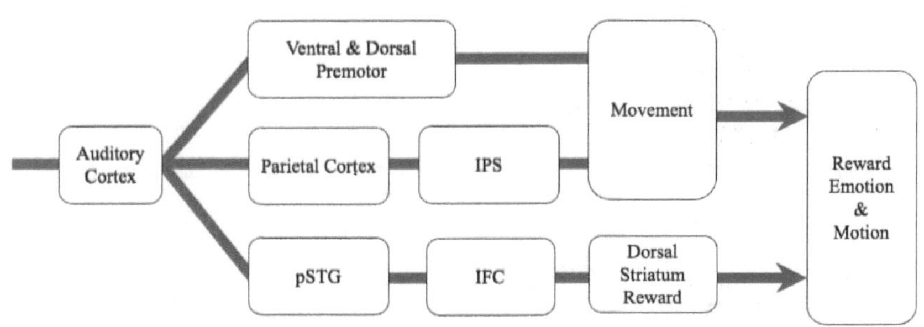

Figure 7.3. This figure illustrates the control system model of the dorsal stream. While feedback and feedforward arrows are not shown to reduce visual clutter, they're present and essential to understanding the neural system and the control system model.

Summary

This chapter traces how music is processed beyond the auditory cortex and follows two main pathways: the auditory ventral stream (AVS) and the auditory dorsal stream (ADS).

The AVS, or the "why" pathway, analyzes the emotional, harmonic, and melodic aspects of music. It detects instruments, recognizes familiar songs, and contributes to emotional and cognitive understanding. It functions through a hierarchy of predictive coding, constantly comparing new music with past experiences.

The ADS, or the "how" pathway, organizes music in timing and space. It tracks rhythm, motion, tempo, and beat, connecting auditory cues to motor responses and coordinating movement. It also connects deeply to the brain's reward systems, especially in social and emotional situations.

Together, the two streams provide a comprehensive framework for understanding music's emotional impact, cognitive intricacies, and motor synchronization. They help explain why music is so deeply embedded in our social lives and emotional experiences.

Theta brainwaves also have a surprising yet profound influence. By aligning with natural Earth frequencies, such as the Schumann resonance, they may indicate a deep evolutionary connection between the human brain and the musical environment.

At both the individual and group levels, musical reward heavily relies on synchronization: internally, between prediction and sensation, and externally, among people. That emotional and social bonding through rhythm and harmony sets the stage for the next chapter.

The descriptions of ADS and AVS processes in this chapter reveal that these regions handle certain similar aspects of music. For instance, they are both activated by rhythmic activities with varying levels of engagement, making the relationship between neural functions/regions and the musical element of rhythm necessarily complex and confusing. It would be simpler if there were a one-to-one correspondence between neural functions or regions and musical elements. Until we better understand musical elements and clearly define neural regions, we may need to accept and live with this complexity and confusion.

Bibliography

1. "Neuroanatomy of language regions of the human brain," Petrides, M. Academic Press. 2014.
2. "From Perception to Pleasure: The Neuroscience of Music and Why We Love It," Zatorre, Robert, *Academic Press*, p. 327
3. "From perception to pleasure: Music and its neural substrates." Zatorre, R. J., & Salimpoor, V. N. *Psychological and Cognitive Sciences*, 2013.
4. "Brain indices of music processing: Non-musicians are musical," Koelsch, S., Gunter, Thomas C., Friederici, D., and Schröger, E., *Journal of Cognitive Neuroscience*, **12** (3), 520-541, 2000.
5. "Toward a neural basis of music perception – A review and updated Model." Koelsch, S., *Frontiers in Psychology*, 2, 2011.
6. "Multistability in auditory stream segregation: a predictive-coding view," Winkler, I., Denham, S., Mill, R., Böhm, T., Bendixen, A., *Phil. Trans. R. Soc. B.* **367** (1591), 2012.
7. "Frequency and frequency modulation share the same predictive encoding mechanisms in human auditory cortex." Stein, Jasmin; von Kriegstein, Katharina; Tabas, Alejandro, 2021 PMC+8arXiv+8arXiv+8
8. "Predictive Coding: A Theoretical and Experimental Review," Millidge, B., Anil K., Buckley, C. L., 2021 DOI: 10.48550/arXiv. 2107.12979

Chapter 8

Reward And Pleasure

The final stage of the "phased-naturalism" sequence (Figure 1.1) links our understanding of how we perceive music (discussed in the previous chapter) to our experience of musical reward, whether pleasure or sadness. After music is processed and decoded mainly through two pathways discussed earlier, the brain begins to interpret it for emotional impact and response. Spectral information like melody is handled by the Auditory Ventral Stream (AVS), while timing and rhythm are managed by the Auditory Dorsal Stream (ADS).

This processed musical information is then sent to several midbrain regions for emotional interpretation, a neural phase that carries out music cognition and reward. This region isn't dedicated only to music or hearing; it processes inputs from all five senses and is part of the highest level in the brain's organizational hierarchy. Because of this change in function and structure, this stage represents a distinct phase in how the brain manages music: one based on emotional evaluation rather than just perception. Additionally, this region functions as a general-purpose system that processes rewards, whether the inputs come from auditory sources or other senses.

The Neural Reward System

Why a Reward System

We know music can thrill us, whether we're listening alone, performing, or moving in sync with a crowd, as discussed earlier in Chapter 2. But what makes it feel so good? Why do we find music enjoyable at all?

The answer lies deep in our evolutionary past. The pleasure we get from music is part of a broader brain system called the reward system, a built-in network designed to help us survive and prosper. Think about how good it feels to enjoy a delicious meal, hear a heartfelt compliment, or experience a comforting touch. That feeling of pleasure isn't random. It's created by a complex neural network, refined over millions of years, to reinforce behaviors that benefit us. While the earliest reasons were simple (stay alive, reproduce), our brains have expanded that system to include more abstract experiences like beauty, learning, and, yes, music.

But this raises a fundamental question: Why does the brain reward us with pleasure in the first place? The answer is elegant. Evolution couldn't afford to wait for conscious decisions in early animals, so instead of intelligence, nature relied on feelings. Creatures, including humans, are wired to feel good when they do something beneficial. That "feel-good" signal is nature's way of saying: "Do this again." For example, when a lioness successfully catches prey, she doesn't just get fed; she also feels satisfied. That satisfaction isn't just a bonus. It's a biological message that encourages her to repeat the behavior, enhancing her survival skills and the chances of passing on her genes.

Humans follow the same basic instinctual rules, but with greater complexity. As our intelligence grew and became more nuanced, so did our emotional and social needs. Still, our reward system serves the same purpose: it promotes behaviors that benefit us, either directly (like eating and resting) or indirectly (like forming relationships, enjoying art, or listening to music). Whether the reward's intent is primitive or advanced, all these actions are driven by one thing: the brain's motivation to reward what helps us thrive.

Learning Through Pleasure

For animals to survive, they must learn and remember how to react to similar situations in the future. This type of learning occurs in two parts. First, the animal must decide *what* to learn. For example, a lion cub needs to develop the skill of hunting. It begins by observing its mother chase prey and memorizing her techniques. The second, and arguably more important, part is *why* the cub is motivated to learn. When it successfully uses that learned skill to catch and eat a gazelle, its brain's reward system kicks in. That surge of satisfaction reinforces the behavior. In other words, learning starts with choosing the right skill and continues by staying motivated to practice and master it, because the brain rewards it.

One can also think of the brain as a pattern-detecting machine. Each time we do something rewarding, it releases a chemical messenger: dopamine. This isn't just the brain's feel-good signal; it's also a learning aid. Dopamine helps connect actions to positive outcomes, shaping future behavior. Over time, we naturally repeat what feels good and avoid what doesn't. That's why rewards are essential in everything from learning to walk as a toddler to riding a bike as a child.

Imagine the moment a child pedals a bike for the first time without falling. They might smile, laugh, or even cheer. That rush of excitement is no accident; it's a dopamine burst that locks in the memory and reinforces the behavior. The brain essentially says, "That was good. Let's do it again." This kind of learning through repeated exposure forms the foundation of statistical learning and predictive coding in cognitive neuroscience. Statistical learning helps the brain recognize patterns. Predictive coding, on the other hand, refines memory by adjusting to new situations, like learning to improve your balance during a bumpier bike ride.

This cycle, which includes action, reward, and memory, is central to how we learn. Whether solving a puzzle, winning a game, or mastering a new dance move, the reward system encourages us to repeat what's worth remembering.

<u>More Than Hunger, Thirst, and Reproduction</u>

The human reward system does much more than satisfy our basic needs like hunger, thirst, and reproduction. It also plays a vital role in maintaining the body's overall balance and well-being. When you feel hungry, that discomfort acts as a warning signal, but what truly motivates you to eat is the pleasure you get from that first bite of food. Similarly, sexual attraction isn't just about reproducing; it's driven by feelings of closeness and emotional connection, which help strengthen bonds and support the continuation of our species.

In short, pleasure is nature's way of motivating us. It keeps us alive and helps ensure we pass on our genes. The reward system acts as a biological nudge, encouraging us not to overlook essential activities like eating, drinking, resting, and bonding. These behaviors are closely linked to neurochemical incentives that keep our bodies functioning and our social lives thriving.

But as intelligent and social beings, humans take this even further. Our brains treat social connection as essential for our well-being. Social rewards, like praise, recognition, and love, can be just as powerful as physical ones. These rewards help build cooperation, deepen trust, and foster strong communities.

We seek belonging not only for emotional comfort but because our brains view social interaction as vital for survival. A smile, a compliment, or even a shared laugh can activate the brain's reward system. Public praise, for example, can release as much dopamine as a paycheck, or sometimes even more. Think about the satisfaction you feel when giving a speech and hearing the audience erupt in applause. I have experienced this emotion personally in a few of my public presentations of my technical work. That's when our reward system lights up because of social connection.

No matter what kind of reward it is—whether physical, emotional, or social—it always triggers chemical reactions in the brain. Think about how hearing your favorite song can give you goosebumps or a chill down your spine when it hits just the right note. That sensation is caused by dopamine.

Deep within your brain, areas like the ventral tegmental area (VTA) and the nucleus accumbens release a mix of feel-good chemicals. These regions are key parts of the reward system, responding not only to primary rewards like food and sex but also to beauty, novelty, music, and even abstract ideas. When dopamine is released, it reinforces the mental connection between an experience and the happiness it causes, encouraging you to seek it again.

While the reward system evolved to reinforce basic instincts, it also encompasses more subtle pleasures unique to humans, especially those that go beyond mere survival.

The Primary and Secondary Rewards

It becomes clear that not all rewards are the same. Generally, there are two kinds of rewards: primary and secondary.

Primary rewards are the most fundamental and biologically ingrained. These are the things directly linked to survival and reproduction, such as food, water, warmth, sex, and physical comfort. Our brains are naturally tuned to these without any learning or teaching.

When we experience a primary reward, ancient regions of the brain become activated, particularly the hypothalamus and the mesolimbic dopamine system, which have been part of our neural makeup for millions of years.

Secondary rewards, on the other hand, are more complex. They don't directly satisfy biological needs, but they *become* rewarding through experience, learning, or cultural significance. These include pleasures derived from beauty, novelty, art, ideas, knowledge, and music. Secondary rewards still activate the brain's dopamine pathways, but they rely more on cognitive processing, memory, and personal or social context. Because they involve more mental effort, the pleasure response may happen more slowly. Interestingly, these rewards can even be triggered *before* the actual stimulus arrives, thanks to the brain's ability to detect patterns and anticipate what's coming from the statistical learning processes.

It becomes clear that music triggers the secondary reward, and a very unique one. Unlike food or sex, music isn't essential for physical survival. Yet, our brains treat it as deeply meaningful. Music activates the same reward pathways as basic needs, especially through the nucleus accumbens, a key center in the brain's pleasure system.

Music also activates the brain's predictive processing systems. We constantly anticipate musical patterns like rhythm, melody, or harmony, and when those expectations are fulfilled (or pleasantly disrupted), the brain provides a small reward because this cycle has occurred before and was stored in memory. This emotional response is often connected to memory, movement, or connection. Additionally, music holds significant social and cultural value: it can reflect shared beliefs, express group identity, and promote feelings of belonging, all of which serve as valuable secondary rewards.

In short, music enables a learned, yet deeply emotional reward. It taps into the brain's universal reward circuitry, making it feel as satisfying, and sometimes even *more* satisfying, than primary rewards like food or physical comfort.

Some researchers suggest that music may act as a bridge between primary and secondary rewards. While it is not essential for physical survival, it may have evolved as a tool for social bonding, emotional regulation, and communication, all of which are vital for human societies. However, this view has a significant flaw. Music, as we recognize it today, likely did not exist until humans learned to combine melodies and rhythms in a structured manner. The evolutionary benefit of bridging the two types of reward could only have appeared after music was invented. Therefore, it's unlikely that the brain developed music-specific reward systems before music even existed. The idea that music evolved as a bridge between reward types doesn't stand up to simple logical scrutiny.

How Are Rewards Triggered?

Have you ever felt a rush of pleasure when biting into a warm, gooey brownie? Or felt chills run down your spine from hearing a particularly thrilling musical moment? Or even experienced joy just from *expecting* something good to happen?

I certainly have. After years of studying how music affects the brain, I've become more sensitive to powerful musical experiences. Take the cannon blasts in Tchaikovsky's 1812 Overture, for example. Just hearing or even anticipating that moment sends shivers down my spine. Before learning about the brain's reward system, that wouldn't have affected me nearly as much. Now, I understand that this thrill is not just an emotional reaction. It's a chemical event deep inside the brain.

That reaction is mainly caused by dopamine, often called the brain's "feel-good" chemical. It plays a key role in how we experience the pleasure of satisfied motivation and learning. All of this is enabled by a neural highway called the mesolimbic pathway, the brain's reward circuit. In fact, this pathway is not a well-defined route; instead, it consists of many connections between the cortices and the midbrain reward system.

The dopaminergic journey begins in a part of the midbrain called the ventral tegmental area (VTA). Situated at the base of the midbrain tegmentum, this area is filled with dopamine-producing neurons that are crucial for reward and motivation.

When the lower parts of the brain—such as the ventromedial prefrontal cortex (vmPFC) or the orbitofrontal cortex (OFC)—detect something surprising or satisfying, they send a signal to the VTA. In response, dopamine is released and travels to the nucleus accumbens, a core pleasure center. VTA also receives signals from the amygdala and releases dopamine to other regions, including the amygdala itself, influencing emotions, memory, and behavior.

Dopamine doesn't just make you feel good; it also teaches your brain. When eating chocolate is pleasurable, your brain saves that information for future reference. Dopamine acts like an emotional bookmark, reinforcing rewarding behaviors. These memories are stored and can be triggered again when a similar situation arises.

The reward system doesn't operate alone. When it's activated, other regions such as the prefrontal cortex also become engaged. This part of the brain assists in thinking ahead, evaluating decisions, and controlling impulses. These systems work together to help you achieve both immediate and long-term goals—even if the reward is delayed. In this way, learning and knowledge acquisition can be rewarding as well, since they satisfy deeper, evolved drives.

By now, it's clear that the brain reacts to satisfaction, relief, and hedonic pleasure. But here's the twist: you don't actually need to *receive* a reward to feel good. Just anticipating a reward—such as smelling popcorn before it's popped or waiting for a message from someone you like—can trigger dopamine release. Sometimes, the anticipation feels even better than the event itself. This type of anticipation is driven by the brain's memory systems, which recognize familiar patterns and predict excitement. This is part of the SLPC (Statistical Learning and Predictive Coding) process in the reward system, which we'll explore more later.

One important thing to remember: the reward system isn't picky. It reacts to experiences from all sensory inputs—sight, sound, taste, smell, and touch. Whether it's food, social interaction, music, or achievement, the reward system adapts to both the physical and emotional experiences. To do that, it requires all incoming signals to be translated into a common neural language—a task handled by the various sensory systems that feed into the reward network.

However, regardless of the source, every reward experience triggers a burst of dopamine. This specific system connects pleasure, motivation, and memory into a unified response.

In the case of music, this system is particularly interesting. The dopaminergic network, including the VTA, nucleus accumbens, and ventral and dorsal striata, helps reinforce patterns that lead to successful predictions or pleasant surprises. For example, dopamine is released in anticipation of a musical climax and upon experiencing the expected resolution.

This dual mechanism, anticipation and experience, directly connects statistical learning and predictive coding to the musical reward experience. Over time, the brain updates its musical expectations based on previous pleasure that influences musical taste, emotional bonds, and even the desire to discover new music.

The Operation of SLPC in the Reward System

What exactly does the statistical learning process learn? Essentially, it learns the entire sequence of events—from the moment a stimulus is received to the ultimate reward that follows. Take this example:

you hear the jingle of an ice cream truck before you see it. Almost immediately, your mouth starts to water and your mood lifts. That reaction is your brain's hidden superpower in action, predicting a reward before it happens. The past experience of hearing that jingle followed by enjoying a sweet treat is ingrained in your memory. That association, learning the connection between the sound and the result, is statistical learning. The storage of this experience for future predictions is predictive coding.

Here's where it gets more intriguing: when your brain predicts a reward and it actually arrives, you get a hit of dopamine—the neuro-chemical that drives pleasure, motivation, and learning. But what if something even *better* than expected happens? More dopamine. And if it's worse? Your brain adjusts, dialing back its future expectations. These differences between expected and actual outcomes are known as reward prediction errors (RPEs); they scale with the magnitude of the difference. They are how your brain constantly fine-tunes its model of the world.

You can think of dopamine neurons as your brain's performance reviewers. They compare what you expected to happen with what actually happened and use that difference to help your brain improve next time. Put simply, your brain is a pattern detector and a prediction machine. Through statistical learning, it figures out the rules of the world. Through predictive coding, it makes informed guesses about what will happen next. When those predictions relate to rewards, dopamine strengthens or adjusts the process. That's how even a simple sound or symbol can become just as pleasurable as the reward itself.

How Primary and Secondary Rewards Differ in the Brain Involvement

Music can activate both primary and secondary rewards, although it tends to evoke stronger responses in the secondary category. Both types involve the brain's reward system, especially the nucleus accumbens, but they do so in slightly different ways, engaging different brain regions and dopamine release patterns.

Primary rewards, such as food or sex, are processed in the mesolimbic dopamine pathway, a central part of the brain's reward system. These rewards trigger rapid and strong releases of dopamine, particularly in areas such as the nucleus accumbens and amygdala.

Secondary rewards, such as money or success, activate the same pathway but also involve higher-level brain regions like the prefrontal cortex and amygdala through PFC. These rewards may lead to slower or more extended dopamine release because they rely more on cognitive processing.

More importantly, primary rewards are linked to biological needs and tend to trigger immediate, strong responses. Secondary rewards are learned through experience and association. They involve more complex mental processes, which can lead to a more nuanced and longer-lasting reward experience. While both types activate the same dopamine system and influence behavior, they differ in how quickly and intensely the brain responds, as well as which brain regions are involved.

Anticipation and Experience Are Different Rewards

The brain doesn't just respond to what is happening; it also tracks when something might occur and how predictable it is. In this sense, the reward system differentiates between two extremes: anticipatory rewards and experiential rewards.

These two types of reward are processed differently. Anticipation is mainly driven by dopamine release in the ventral striatum and is linked to a motivational state often called "wanting." In contrast, experience relates to actually receiving the reward and results in "liking," the hedonic pleasure we feel at the moment.

These two responses, wanting and liking, are distinct yet interconnected. Wanting is the internal desire or urge to get a reward. It often happens unconsciously and is based in the brain's dopaminergic systems. Liking is the actual pleasure or enjoyment felt during the reward. It's the personal feeling of satisfaction that shows the reward was worth it.

To illustrate: when you anticipate a positive experience, like seeing a friend after a long time, your brain's reward system activates before anything occurs. This anticipation increases dopamine levels in the ventral striatum, sparking arousal, excitement, and even improved well-being. This feeling is anticipatory and "wanted." Once the experience actually takes place, such as hugging your friend, the brain's reward system becomes more active. More dopamine is released, and the moment is stored as a pleasurable memory. This stage involves "liking" and helps strengthen the emotional memory.

The main difference between these two types of reward lies in timing. Anticipation happens before the reward and is associated with building expectation and motivation. Experience occurs during or after the reward and involves actual pleasure or satisfaction. In musical terms, studies show that dopamine levels can spike a few seconds before the finale of a musical piece, indicating that the brain is already rewarding the listener in anticipation of the climactic moment. The a priori knowledge of the impending finale is stored in the memory through the statistical learning process.

Musical Pathway Connecting Reward System

The previous section describes how music connects to the brain's reward system. Now, let's explore the specific neural pathways that route this sequence of neural signals.

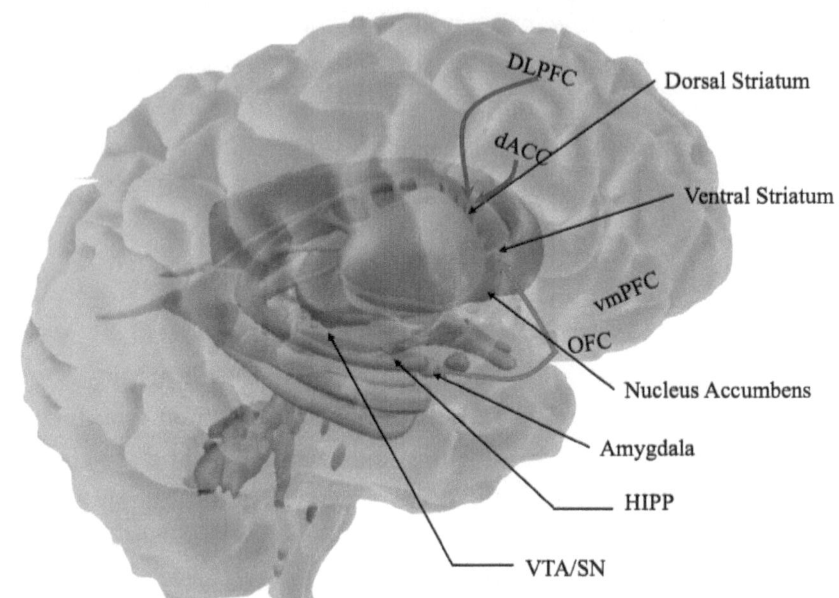

Figure 8.1. An illustration of the inside of the right hemisphere, highlighting the brain regions involved in processing musical reward. The blue and red lines represent the connections to the reward areas of the dorsal and ventral striata. The brain outline is based on the 3D atlas from Neurotorium, https://neurotorium.org/ tool/brain-atlas/.

Figure 8.1 shows a 3D view looking into the right hemisphere, highlighting the brain regions involved in processing musical reward. This figure also builds upon Figure 7.1, where the ventral and dorsal streams are depicted as blue (dorsal stream) and red (ventral stream) lines, respectively. These connections receive the analyzed signals generated in the ADS and AVS, and the lines extend from the outer brain to the midbrain, i.e., the reward areas of the dorsal and ventral striata. Figure 8.1's blue and red lines continue from the blue and red lines in Figure 7.1.

Figure 8.2 plots the logical flow of the signals and the pathways, where the blue and red colors also follow the convention used in Figures 7.1 and 8.1.

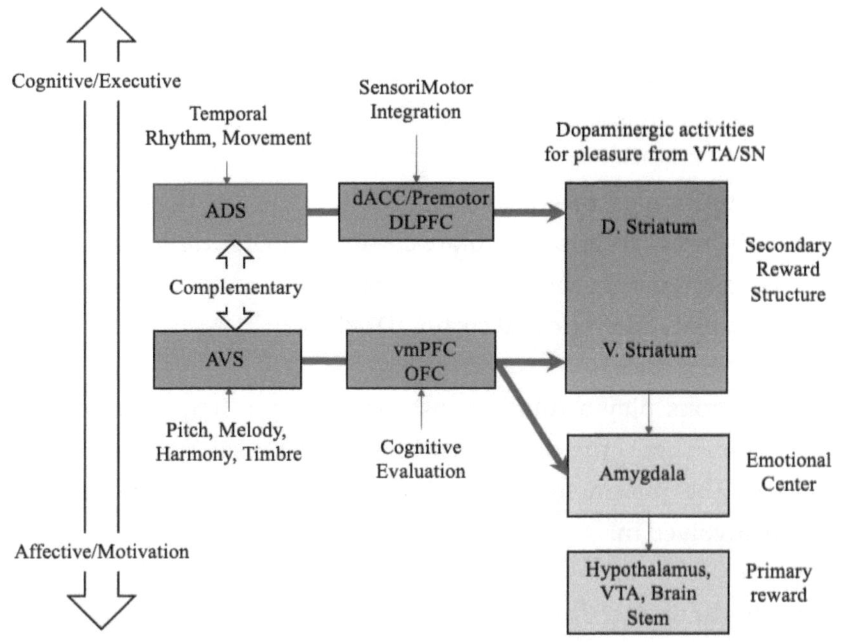

Figure 8.2. This figure shows the flows of AVS and ADS signals into the reward system.

As discussed in the previous chapter, the brain processes music via two complementary auditory pathways. The auditory dorsal stream, often called the "how" pathway, handles temporal processing, rhythm perception, and integrates music with movement. The auditory ventral stream, or "what" pathway, focuses on decoding structural elements of music, such as pitch, harmony, and timbre.

Music is a uniquely auditory stimulus that triggers strong emotional and motivational responses. The brain processes it in multiple ways—activating perceptual, cognitive, and emotional mechanisms that connect directly to both primary and secondary reward systems. Then the ventral and dorsal auditory streams work together to direct musical information to the brain's pleasure centers.

The Dorsal Stream: Linking Rhythm and Movement to Reward

The dorsal stream extends from the primary auditory cortex (A1) to multiple brain regions.

- Posterior parietal cortex
- The dorsal anterior cingulate cortex (dACC) is tightly connected with the premotor cortex
- Dorsolateral prefrontal cortex (DLPFC)

These regions play a role in sensorimotor coordination and attention. From there, the signals travel to the dorsal striatum, which includes the putamen and caudate nucleus. These structures are deeply involved in:

- Rhythmic entrainment
- Motor learning
- Habitual response selection

In short, when you tap your foot to a beat or get up to dance, you're activating this dorsal stream-to-striatum loop, which connects movement to musical reward. The rhythm becomes not just something you hear, but something you feel and act on. (Figure 8.2 illustrates these key regions and how the information flows through them.)

The Ventral Stream: Mapping Musical Identity to Emotion and Value

This stream also originates in the primary auditory cortex (A1) but travels ventrally, passing through the anterior superior temporal gyrus to reach higher-order brain regions such as:

- Ventromedial prefrontal cortex (vmPFC)
- Orbitofrontal cortex (OFC)

The OFC is vital—it's a central hub in the secondary reward system. It assigns subjective value to sensory stimuli. When music processed through the ventral stream reaches the OFC, it can gain hedonic value, especially if the music is emotionally resonant or aesthetically pleasing.

This information is then transmitted to the ventral striatum, and particularly to the nucleus accumbens, which is key to:

- Musical pleasure
- Reward anticipation
- Emotional resonance

The amygdala, known for its role in emotional memory and valence processing, also receives input from auditory and prefrontal areas. It contributes to the emotional intensity of musical experiences and connects both to the hypothalamus and ventral striatum. These links bridge the primary reward system (which governs physiological responses) and the secondary reward system (which assigns learned emotional value).

<u>Integrating the Primary and Secondary Reward</u>

The primary reward system, which includes the hypothalamus, ventral tegmental area (VTA), and brainstem autonomic centers, supports strong physiological and emotional reactions to music. These responses can be so intense that they cause goosebumps, chills, or even tears. They are usually triggered indirectly through pathways that connect the auditory system to the amygdala and prefrontal cortex.

Meanwhile, the secondary reward system, made up of cortico-striatal circuits—particularly the ventral and dorsal striatum—manages the learned, cognitive, and emotional aspects of music appreciation. This includes:

- Musical preferences
- Expectation of musical patterns
- Reinforcement through repetition

In summary, musical reward results from the combined activation by both the dorsal and ventral auditory streams. These pathways direct musical information to emotional, cognitive, and motor systems, which then integrate at both the primary physiological and secondary emotional-cognitive reward centers. This complex interaction helps explain why music can move us so deeply, shape our behavior, and bring people together.

The Musical SLPC in the Reward System

Previously, we examined how Statistical Learning and Predictive Coding (SLPC) function in the reward system without focusing on any specificity. These mechanisms are sensory agnostic, meaning they work regardless of whether the stimulus is visual, tactile, or auditory.

But music adds something special. In the case of musical stimuli, the reward system doesn't just respond; it judges. It assesses musical features based on past experiences and expectations. This is where SLPC becomes essential. The brain creates internal models of musical structure based on what we've heard before. These models allow the brain to predict what will come next in a melody or rhythm. When the actual music either matches or intentionally breaks these predictions, a reward response is triggered. When music has moderate complexity, these prediction errors tend to be particularly satisfying, leading to enjoyment and a desire to move with the music.

Music also differs from other types of stimuli because it is highly repetitive. More than 50% of Western music, for example, depends on repeating melodies, harmonies, rhythms, and patterns. This repetition helps musical stimuli and their related rewards become deeply stored in memory, especially in areas like the parahippocampal region, which boosts reward signals.

But musical reward also varies based on individual personality. Someone adventurous might enjoy more surprising or unconventional musical elements because these create higher levels of predictive error, which the brain finds exciting.

Similarly, a person's musical training or exposure can influence their reward response. It's like solving a math problem: if the problem is too complex, you might give up, and no reward is experienced. If it's too easy, there's no challenge, so again, no substantial reward. The ideal level is in the middle. This inverted U-shaped curve, where moderate complexity is most rewarding, is well-supported by studies on musical rhythm, tonal tension, and information content.

Why does music make us want to move? It's because it's internalized or stored in the motor regions of the brain. Strong predictive errors, moments of surprise, can trigger pleasure, dopamine release, and a desire to move. The synchronization of melody and rhythm with body movement isn't just enjoyable; it also has powerful social effects, which we'll explore further in the following sections.

In summary, musical rewards come from a detailed analysis of auditory input. This input is processed by brain regions such as the vmPFC, OFC, dlPFC, and sensorimotor systems, ultimately resulting in a secondary reward experience. While there might be a small primary reward contribution via the amygdala, the main process occurs through dopaminergic release in the midbrain—specifically the VTA and brainstem.

Control System Modeling of Musical Reward System

To better understand how SLPC operates within the reward system, we can again parallel it to the engineering control system. The reward system's control system model utilizes feedback and feedforward loops to process information and adjust responses based on experience.

Figure 8.3 describes this model. In this system, pleasure and movement are the visible outcomes of the reward mechanism. These responses are influenced by how the current stimulus compares to stored memories. That comparison, between expectation and actual result, drives both anticipatory and experiential rewards.

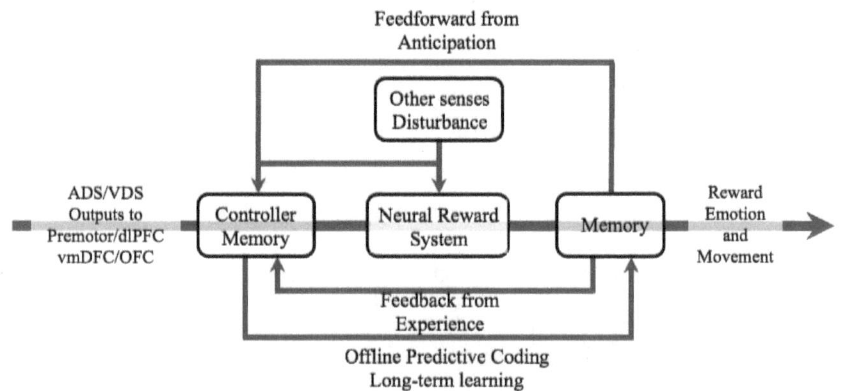

Figure 8.3. Figure 8.3 describes the processes of the musical reward in a control system model.

Here's an example: imagine going to a concert where Beethoven's Fifth Symphony is being played. You expect the final note to be a bold C major chord, played fortissimo and with perfect timing. As the music builds toward this climax, your brain anticipates the moment and starts releasing dopamine, even before the final note is played. This anticipatory pleasure comes from the feedforward pathway, which is indicated on the top line of the control system diagram.

But what if the conductor decides to delay that final chord by just a quarter of a second for dramatic effect? The unexpected pause surprises the audience and creates an extra rush of pleasure. This kind of response is experiential; it occurs after the stimulus and is processed through the feedback pathway at the bottom of the system model.

Now, here's where things get even more interesting. After experiencing it, the brain's offline feedforward mechanism updates its internal model. It stores details about the variation in timing and the related emotional response, integrating them into memory. The next time you hear the same piece, especially under the same conductor, your brain may anticipate the delay and adjust its expectation accordingly.

This model, from stimulus to updated memory, demonstrates the SLPC loop in action. Whether feedforward (online or offline) or feedback, each signal helps modify the brain's memory model, reinforcing or reshaping expectations based on past musical encounters.

In short, the brain's reward system functions like a self-regulating control mechanism, continually comparing input to past experiences, adjusting predictions, and fine-tuning responses to maximize pleasure and meaning.

The Pleasure of Synchronization

Auditory Dorsal Stream and Movement

Chapter 7 had briefly described how the auditory dorsal stream processes movement and synchronization. We dive deeper in this section.

Movement, especially dance, is an integral part of the musical experience. Music causes motion by activating a connection between the auditory dorsal stream and the premotor cortex through a process called sensorimotor integration. The dorsal auditory stream processes auditory information, especially rhythmic patterns in music, and transmits signals to the premotor cortex, which handles planning and choosing movements. This connection enables music to evoke both automatic and intentional physical responses. For instance, the brain might translate a steady beat into the motor command to tap your foot or execute a complex dance move.

The auditory dorsal stream also helps us process "where" sounds occur in space and how they relate to movement. It manages the rhythmic structure of sound and sends that information to the premotor cortex, which also receives inputs from other sensory and motor regions.

This interaction enables the seamless integration of auditory input and motor planning, which is crucial for activities such as dancing to a beat or clapping in time. Some movements, such as tapping your foot to the rhythm, happen automatically. Others, like following choreography, are intentional. But both are made possible by this neural connection.

The temporal structure of music, including its tempo, beat, and rhythm, is essential to this process. The dorsal stream interprets these patterns, and the premotor cortex converts them into coordinated physical movements.

Dopamine Reward and Movements

Movement and music are connected not only through motor circuits but also via the dopaminergic reward system. When we move to music, our brain releases dopamine, especially in the striatal system, reinforcing both the movement and the enjoyment.

This process creates a positive feedback loop (again, not the microscopic one): the rhythm prompts movement; the movement feels enjoyable; and the dopamine release motivates us to repeat it, part of the statistical learning. The brain's ability to predict musical events and synchronize its motor responses is deeply satisfying. When the timing is perfect or pleasantly unexpected, it produces a surge of pleasure. You've probably felt this yourself: tapping your foot, swaying to a tune, or dancing to a beat. These actions cause dopamine to be released, forming a loop through hearing, movement, and pleasure. That's why moving to music feels so good, and why we keep doing it.

While physical movement enhances the dopamine response, it's important to remember that dopamine also supports emotional, memory, and cognitive responses to music. In other words, even without moving, music can still deeply engage our brains.

Synchronization and Reward

Synchronization and syncopation, two rhythmic features in music, can activate reward systems through multiple mechanisms. Both stimulate the dopaminergic network and engage motor systems, reinforcing a sense of pleasure and a desire to move. Rhythms of medium complexity, especially those with mild syncopation, are particularly effective. They activate areas like the ventral striatum and nucleus accumbens, creating the enjoyable sensation often called the "groove". Even without physical movement, these rhythms activate motor regions in the brain. This can lead to a natural urge to synchronize with the beat, even if you don't consciously respond to it. Syncopation, which introduces rhythmic surprises by shifting beats to unexpected positions, adds novelty. The brain responds to this by adjusting its predictions and rewarding the effort with further dopamine release because syncopation is more complex to follow and brings more satisfaction when successful. The relationship between rhythm and reward follows an inverted U-curve: rhythms that are too simple or too complex are less rewarding. The brain is most pleased by a balance of predictability and surprise.

In social contexts, this synchronization becomes even more rewarding. Sharing timing and dynamics with others helps transfer emotional states, creating a strong sense of unity and shared enjoyment.

Essentially, syncing with music, whether regular or syncopated, engages both the reward and motor systems, creating a pleasurable experience with social and emotional aspects.

How Does Synchronization Spread Over Group Activities

Group synchronization, like what happens at concerts, relies on a mix of auditory cues, emotional contagion, and physiological alignment. People don't just hear the beat—they move, feel, and sometimes even breathe in sync or harmony.

- Auditory cues provide the rhythmic structure for coordination. The beat and meter of the music enable people to time their movements together.
- Emotional sharing enhances this effect. Live music often sparks strong feelings, and these shared emotions deepen the sense of connection. Research shows that audience members can synchronize heart rate, breathing, and skin conductance—physiological signs of alignment with the music and one another.
- Movement synchrony has also been observed, where concertgoers naturally fall into rhythm with one another, further strengthening group cohesion.

There are also different types of synchronization:

- Induction synchrony, where the music induces collective movement.
- Interaction synchrony, where performers—such as musicians and conductors—coordinate with one another.

Endorphins released during synchronized movements also boost feelings of bonding and pleasure, increasing the sense of belonging.

Musical Synchronization and Social Bonding

Musical synchronization, especially when it involves mild syncopation, plays a significant role in social bonding and emotional connection. Whether it's clapping in rhythm or dancing together, coordinated movement has been shown to boost positive social feelings and foster a sense of belonging. Syncopation, which emphasizes unexpected beats, can make music feel more enjoyable and even happier. This emotional boost improves our social experience.

Research shows that syncing with others during musical activities helps create shared intentions and enhances interpersonal connections. It results in increased feelings of closeness and mutual understanding.

Music's ability to entrain behavior across cultures emphasizes its role in social communication. Group synchronization fosters prosocial behaviors like empathy, cooperation, and trust. Brain regions involved in motor coordination, reward, and social cognition are all active during these experiences.

While everyone can benefit, the effect depends on individual preference and musical experience. People who enjoy dancing or rhythmic, clear music may experience the effects more intensely.

Musical Synchronized Movement and Chaos

Chaos theory, a branch of physics, studies unpredictable behavior in systems that usually seem stable. It examines how nonlinear dynamics can produce surprising yet patterned results. A famous experiment by the Ikeguchi Laboratory in Japan used 100 metronomes to demonstrate this nonlinear behavior. Each metronome is tuned to the same frequency, but it is not synchronized with the others at the exact phase on the same platform. However, over time, all metronomes gradually evolved into a fully synchronized collective movement. This was achieved through the nonlinear interaction between them because they all sit on a freely moving platform.

This process is captured in a video: https://www.youtube.com/watch?v=suxu1bmPm2g. In the video, it appears as though the metronomes are "communicating"—the ones oscillate at more advanced phases retard, the ones oscillate at retarded phase advance. In reality, though, they're all responding to the same shared foundation—the platform beneath them. As the platform reacts to each metronome's movement, it adjusts, gently nudging them into alignment in phase. Eventually, the system finds a low-energy, stable state, a process similar to entrainment in neuroscience.

The synchronization of musical rhythm in a concert behaves in a remarkably similar way. In this analogy:

- The metronomes represent individual concertgoers
- The shared platform symbolizes the neural and emotional connection among people
- The "communication" occurs not through physical contact, but through neurological synchronization

This analogy came to me during a 2022 Beach Boys concert. As the band played under the night sky, the audience raised their phones, swaying in sync with the music. It was a moment of profound, unspoken connection, with entrainment in action.

Both systems, metronomes and music, rely on perceiving tempo and rhythm. Music introduces additional complexity, such as polyrhythms, but the basic principle remains the same. Rhythms are processed in the auditory dorsal stream, which then communicates with the premotor cortex to synchronize movement.

While the metronome experiment is purely physical, musical synchronization has social and emotional effects. It promotes cohesion, a shared identity, and group experience, all without direct contact. Instead, the interaction is mediated by dopamine, making it biologically rewarding.

Music's capacity to produce powerful states of synchrony, even among strangers, greatly enhances its enjoyable effects and has even been suggested as a therapeutic tool. Religious and spiritual traditions across cultures often use music to induce these states, sometimes pushing them toward trance or expanded consciousness. Musical timing reflects our social environment and, in turn, shapes how we understand and connect with others.

Of course, we still face the evolutionary question: Has this kind of social synchronization ever benefited human survival? The answer remains elusive. But we can look to fireflies for a clue. Their synchronized flashing is believed to help in mate recognition, boosting their reproductive success. Could musical synchronization have similar long-term benefits for humans in light of the speculation by Smith, Hume, and Schopenhauer (Chapter 1)? Maybe.

However, at this point, the evidence remains speculative. What is clear, however, is that synchronized music experiences have a powerful emotional and social impact, one that continues to influence our lives in ways that are both visible and deeply felt.

Summary

Once the auditory cortex processes music and it flows through the auditory ventral and dorsal streams, a transformation occurs: the brain's reward system takes over. At this point, information from the auditory system is fused with inputs from other sensory channels, like vision, smell, and taste, to be evaluated by the sensory-agnostic reward system. This reward system doesn't care which sense triggered the input; instead, it assesses the importance of the experience and decides how the brain and body should respond.

Over millions of years, this system has evolved to support human well-being, ensuring we engage in behaviors that promote survival, emotional stability, and social bonds. Once triggered, it produces responses that range from satisfaction and pleasure to physical movement, especially when influenced by rhythm and musical structure.

Humans' brains recognize two main types of rewards. Primary rewards are instinctive and biological, like food, water, and reproduction. They are essential for survival and hardwired into our brains. Secondary rewards are more nuanced. They develop through learning, culture, and experience, and often include abstract pleasures such as beauty, art, knowledge, and, of course, music.

Music falls squarely into the secondary reward category. It's not essential for survival, yet it stimulates deep pleasure through our evolved reward circuitry. Listening to music triggers the release of dopamine, the brain's "feel-good" neurotransmitter, suggesting that music is primarily tied to the brain's "wanting" system, the emotional, reflective kind of reward, rather than the more primal "liking" system associated with instinctual urges.

This distinction between wanting and liking is essential. Wanting is often more emotional and cognitively processed, linked with anticipation, surprise, and interpretation. This reward plays a bigger role in anticipatory rewards, like the release of dopamine occurs before the music reaches a climax (such as the buildup before a powerful chorus), while experiential rewards happen during or after a fulfilling musical moment.

These complex reward mechanisms are guided by the principles of Statistical Learning and Predictive Coding (SLPC). As shown in the control system model (Figure 8.3), the brain depends on past experiences and learned patterns to anticipate and respond to musical events. This cycle allows the brain to refine its expectations over time, making music more enjoyable the more we understand it.

Finally, music isn't just something we hear; it's something we feel and often move to. This movement originates from the Auditory

Dorsal Stream (ADS), which connects to the premotor cortex. While the Ventral Stream is more involved in decoding the identity of musical features like melody and harmony, it's the dorsal stream that controls rhythm perception and synchronization, especially the kind that makes us tap our feet or dance.

When we physically synchronize to a beat, especially in social settings like concerts, an extra layer of reward is triggered. Matching our movements with music—particularly when navigating syncopated (off-beat) rhythms—can stimulate increased dopamine release. However, if syncopation becomes too complex and hard to follow, enjoyment declines. This illustrates the well-known inverted-U curve of musical complexity: our brains aim for a balance between predictability and surprise.

This phenomenon of synchronization fosters a shared sense of time and rhythm among groups of people. The result is a strong feeling of camaraderie, emotional connection, and even ideological unity, which is why music has long been used in everything from religious rituals to political movements.

Music's ability to synchronize not just our bodies but also our minds and emotions explains why it is so rewarding and remains one of humanity's most cherished and powerful experiences.

Bibliography

1. "Anatomically distinct dopamine release during anticipation and experience of peak emotion to music," Salimpoor, V. N., Benovoy, M., Larcher, K., Dagher, A., & Zatorre, R. J. *Nature Neuroscience,* **14**(2), 257–262, (2011). https://doi.org/10.1038/nn.2726
2. "From perception to pleasure: Music and its neural substrates," Zatorre, R. J., & Salimpoor, V. N., *Proceedings of the National Academy of Sciences,* 110(Supplement 2),

10430–10437, (2013). https://doi.org/10.1073/pnas.
1301228110

3. "Brain correlates of music-evoked emotions," Koelsch, S.,
    Nature Reviews Neuroscience, **15**(3), 170–180 (2014). https://
    doi.org/10.1038/nrn3666

4. "What is the role of dopamine in reward: Hedonic impact,
    reward learning, or incentive salience?" Berridge, K. C., &
    Robinson, T. E., *Brain Research Reviews*, **28**(3), 309–369,
    (1998). https://doi.org/10.1016/S0165-0173(98)00019-8

5. "A neural substrate of prediction and reward," Schultz, W.,
    Dayan, P., & Montague, P. R. *Science*, **275**(5306), 1593–1599,
    (1997). https://doi.org/10.1126/science.275.5306.1593

6. "The reward circuit: Linking primate anatomy and human
    imaging," Haber, S. N., & Knutson, B.
    Neuropsychopharmacology, **35**(1), 4–26, (2010).

7. https://doi.org/10.1038/npp.2009.129

8. "Maps and streams in the auditory cortex: Nonhuman
    primates illuminate human speech processing,"
    Rauschecker, J. P., & Scott, S. K., *Nature Neuroscience*, **12**(6),
    718–724. (2009). https://doi.org/10.1038/nn.2331

9. "Musical reward: From musical preferences to emotional
    communication," Loui, P., & Schlaug, G., *Frontiers in
    Psychology*, **3**, 495, (2012). https://doi.org/10.3389/fpsyg.
    2012.00495

10. "A theory of cortical responses," Friston, K. *Philosophical
    Transactions of the Royal Society B: Biological Sciences*,
    360(1456), 815–836, (2005).

11. https://doi.org/10.1098/rstb.2005.1622

12. "Auditory expectation: The information dynamics of music
    perception and cognition,"

13. Pearce, M. T., & Wiggins, G. A., *Topics in Cognitive Science*,
    4(4), 625–652, 2012)

14. https://doi.org/10.1111/j.1756-8765.2012.01214.x

15. "Sensorimotor coupling in music and the psychology of the groove," Janata, P., Tomic, S. T., & Haberman, J. M. *Journal of Experimental Psychology: General,* **141**(1), 54–75, (2012). https://doi.org/10.1037/a0024208

16. "Feeling the beat: Movement influences infant rhythm perception," Phillips-Silver, J., & Trainor, L. J. *Science,* **308**(5727), 1430 (2005). https://doi.org/10.1126/science.1110922

17. "Rhythm and beat perception in motor areas of the brain," Grahn, J. A., & Brett, M. *Journal of Cognitive Neuroscience,* **19**(5), 893–906, (2007).

18. https://doi.org/10.1162/jocn.2007.19.5.893

Chapter 9

Quality of Harmonies

What Harmonies Please Our Reward System

Music is a human invention that draws on the full spectrum of naturally occurring elements, such as sound, voice, rhythm, instrumental tones, structure, and language, to create something uniquely expressive and meaningful. Western music includes one additional element: harmony. The development of harmony has a relatively brief history compared to other musical elements. Although different cultures, such as the Chinese, have also developed ensemble music using multiple types of instruments and tones, the concept of harmony has not been fully crystallized for further development. The harmony in Western music is unique in its wide application and sophisticated development. It began with early polyphony and continued into the Renaissance. By the time of the Baroque period (approximately 1600–1750), functional harmony had become a foundational element of musical composition. Today, harmony is almost ubiquitously present in modern music. Most music pieces in all cultures use it to enhance emotional expressiveness.

In previous chapters, we traced the journey of music from the ears through cognition and finally into the realm of pleasure, both personal and societal, completing what we have called the whole sequence of phased naturalism. Throughout that exploration, we assumed that each component of music originated from sounds, whether vocal or instrumental, produced by naturally occurring materials.

However, one aspect of harmony challenges this idea. Unlike rhythm or melody, harmony is not naturally produced by a single instrument or voice. Instead, harmony is created artificially by combining multiple tones played or sung at the same time, in the form of chords, whether dyads or more complex structures. They differ from naturally occurring sounds because they cannot be described solely by fundamentals and harmonic octaves (Chapters 3 and 7).

The most interesting follow-up question, then, is: if harmony doesn't happen naturally, how do human brains distinguish between pleasant and unpleasant harmonies, i.e., between consonant and dissonant sounds? Interestingly, evidence shows that the brainstem, one of the earliest relay points in our auditory processing chain, can already differentiate between consonant and dissonant tones. As suggested in Chapter 6, this may be because the statistical learning and predictive coding (SLPC) processes, both feedforward and feedback, span from the brain's highest cognitive centers to its low-level neural structures, such as the brainstem. But if the consonance perception is supposedly created artificially during written history, hardly any time for brain evolution to happen, how can the lowest primal level neural structure be able to discern its quality?

There are, however, alternative perspectives emerging from fields that seem far removed from neuroscience. These studies suggest that perceiving harmony might be encoded in the physics of how a few distinct tones interact, especially when the combined tones are not naturally produced.

This chapter deviates from the phased naturalism framework to explore this question: although the human brain is capable of discerning the quality of harmony to be consonant (pleasant) or dissonant (unpleasant), where and how that judgement is made. Is it a neural response or a more fundamental physical process?

Long before neuroscience emerged, scholars were already fascinated with the origins of consonance and dissonance, a curiosity that goes back to the 5th century BCE. But before exploring the tone quality debate further, it's important to recognize that consonant and dissonant feelings are real, measurable phenomena, perhaps as fundamental as our innate musical instincts and part of the musicality, making them worthy of scientific investigation.

Figure 9.1 shows how people rate the pleasantness of two tones played together at different intervals, from one to eleven semitones.

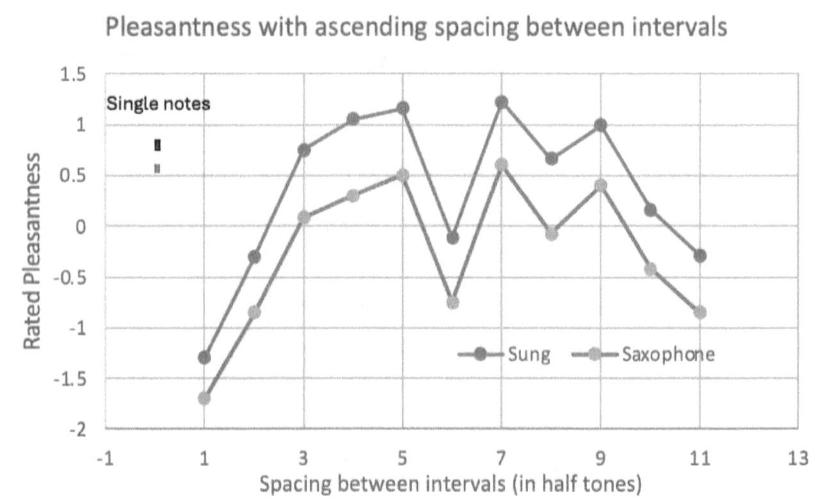

Figure 9.1. *This figure illustrates how people rate the pleasantness of two pure tones played together across varying intervals, from one to eleven semitones. The figure provides a visualization that shows how pleasantness varies with the spacing between tones. The blue line represents the chords sung by humans, while the red line shows the same chords played on a pair of saxophones.*

The figure visualizes how pleasantness changes with the distance between the tones. The blue line represents the chords sung by humans, while the red line shows the same chords played on a pair of saxophones.

Notably, the least pleasant combination occurs when the two tones are just one semitone apart. Pleasantness steadily increases from that point until the interval reaches six semitones, at which point it drops sharply. This six-semitone interval is known as the tritone, a particularly jarring sound that lies three whole tones away from the octave. Historically referred to as "The Devil's Interval," it was avoided for centuries by composers and their students. Still, as we'll see later, this "diabolical" tone has its uses in modern Western music, especially when embedded within larger harmonic contexts. The most pleasant interval, as preferred by participants in the study, was seven semitones apart, a combination commonly known as the perfect fifth.

This large-scale experiment involved approximately 250,000 participants, all of whom were tested in a way that eliminated any influence of formal musical training. This ensures that the preferences observed were not shaped by education or exposure but instead reflect a deep-rooted tendency. It is also notable that, under these experimental conditions, the human voice was rated more pleasant than musical instruments, with a score of 0.85 for the human voice compared to 0.55 for saxophones.

There are independent studies showing that the preference for consonant tones begins very early in humans' lives. Remarkably, even a five-month-old chimpanzee raised by humans preferred consonant music. Using a computerized setup, the infant chimpanzee consistently chose artificially produced consonant sounds for longer durations than dissonant ones. This suggests that the preference for consonance may not be uniquely human but rather a biological trait shared with some primates. However, the robustness of this experiment is slightly affected by the fact that the chimp was raised by humans. Nonetheless, it remains an interesting data point.

We are now on firmer ground and believe that harmony does have a qualitative merit and can be detected as neurological responses. As such, the fact that the harmony can elicit primates' innate emotional cognition suggests that this feeling might be more fundamental than high-level neural processes.

This chapter examines the consonance level of chords from a purely physical perspective, aiming to connect physicists and music researchers, as well as the worlds of physics and music. We will show how rigorous physical studies, using simplified and realistic models, reach conclusions about tone pleasantness without relying on neuroscience.

This chapter begins by characterizing human and instrumental sounds in both the temporal domain (how sound waves change over time) and the spectral domain (the audio frequencies they contain). We show that most naturally occurring "single" tones, whether vocal or instrumental, are multi-toned, consisting of a fundamental and harmonics. Spectral graphs and waveform analyses (obtained using iPad apps at home) reveal that these tones usually have a dominant frequency, supported by additional harmonics spaced at whole-number multiples, a hallmark of consonance.

Next, we briefly review how neural impulses are generated in the inner ear to set the stage for our physical analysis. (The microscopic mechanisms were already discussed in Chapter 4.) The rest of the chapter is devoted to correlating pleasant tones with regular neural firing and unpleasant tones with chaotic firing patterns. We approach this from two different angles and methodologies, both of which have equally strong physical bases, although their fundamental assumptions differ.

First, we observe that consonant tones share a physical characteristic: they involve combinations of two or more pure tones with simple frequency ratios. These simple ratios explain why the auditory system can lock onto a rhythm of impulses that "sync up" in predictable cycles. Second, by using probabilistic modeling for the noisy environment in which neural impulses are immersed, we examine how the regularity of neural firing correlates with these same simple ratios. When tones align this way, neural activity becomes more predictable, and the sensation becomes more pleasant.

The analysis clearly demonstrates that both methods arrive at the same conclusion: tones arranged according to Farey sequences on audio frequencies tend to align with our perception of consonance. In other words, when played or sung together, these tones create harmonious relationships that the brain naturally finds pleasing, regardless of where the judgment occurs in the brain.

We'll conclude by tying these findings to familiar concepts about how we judge the quality of harmony, giving us a more complete understanding of why certain musical combinations sound "right" and others do not.

What is in a Tone

Single-syllable tones, whether produced by the human voice or musical instruments, can be scientifically represented in two complementary ways: the temporal domain and the spectral, or frequency, domain. This dual representation aligns with the well-established time and place analysis of hearing in the inner ear (see Chapter 5). In scientific terms, tones are described either by their time-dependent waveforms (how air pressure changes over time) or by their constituent frequencies (the specific pitches that make up the sound). Together, these two views provide a comprehensive picture of a tone's structure.

But how do we visualize sound? Scientific tools now enable us to graph air pressure changes over time, producing waveform graphs that map the actual air disturbance as it occurs in time. These waveforms are rarely repetitive or straightforward, contrary to what one might expect. They often contain hidden layers, subtle frequencies embedded within the primary waveform. To uncover these components, we utilize a mathematical tool known as the Fast Fourier Transform (FFT). This technique breaks down complex waveforms into their constituent frequency components, presenting them as spectral graphs that show the strength of each frequency across its frequency range. In simple terms, waveform graphs represent how the air pressure varies over time (the time domain), and spectral graphs reveal the specific frequencies present in that sound (the frequency domain). This description of sound might be a bit too technical, but the idea of complementarity between time and frequency is as old as Pythagoras. FFT has also been around for about 200 years since Joseph Fourier invented it. It has become so widespread that an iPad app is offered as one of the free downloads used in this chapter.

When a spectral graph displays a tone consisting of only one frequency, we refer to it as a pure tone (PT). This idealized sound rarely occurs in nature but is often used in psychoacoustic experiments to study hearing. A pure tone produces a smooth, periodic waveform, with its frequency matching the single spike shown in the spectral graph. For example, Figures 9.2a (left) and 9.2b (right) display the waveform and spectral graph of a computer-generated pure tone at 1,046 Hz (corresponding to the C6 piano key), obtained using the Apple application.

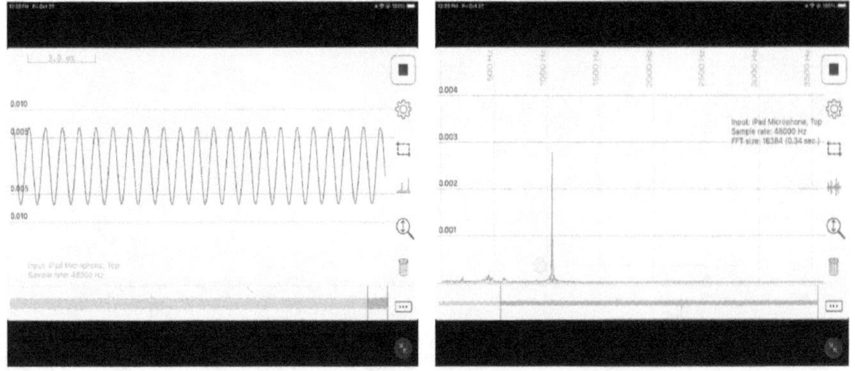

Figure 9.2. Figure 9.2a displays the waveform and Figure 9.2b the spectral graph of a computer-generated pure tone at 1,046 Hz (corresponding to the C6 piano key), obtained using the Apple application program SoundSpectrum.

The real-world tones, whether vocal or instrumental, are rarely pure. They usually have a dominant, or fundamental, frequency—the note we hear—accompanied by several weaker tones at multiples of the fundamental, known as harmonics. I recorded myself singing the C4 note (261.6 Hz) and analyzed both the waveform and spectral graph, as shown in Figures 9.3a (left) and 9.3b (right). Although my voice is untrained, it represents a natural and typical rendition of the C4 pitch. The spectral graph reveals additional frequencies at two, three, and four times the C4 note; these are harmonics of the fundamental tone. This indicates that my singing of C4 includes at least three harmonics and covers over four octaves. Notably, the component two octaves above C4 (around 1,000 Hz) is very close to C6, as expected, indicating that my singing of C4 is on the note.

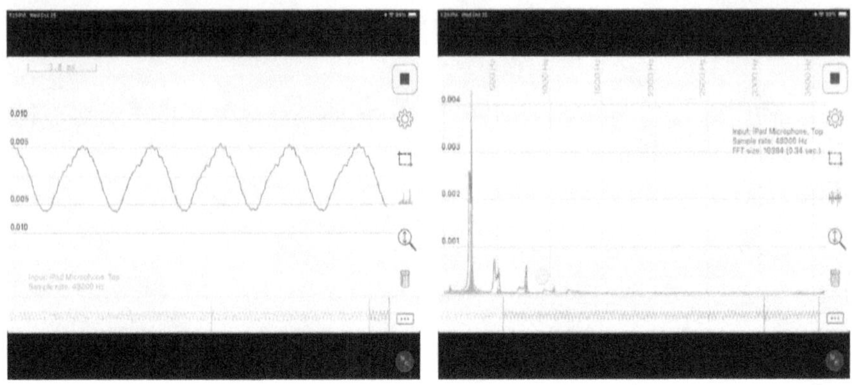

Figure 9.3. A waveform and spectral graph of my voice singing the note at 261.6 Hz.

Figures 9.4a and 9.4b display the waveform and spectral graph of a C4 key on the piano. Like my voice, the piano tone contains multiple frequency components. The waveforms for both my singing and the piano's C4 note show additional layers of periodic waveforms beyond the primary frequency. These layers, visible in both time and frequency domains, create the sonic fingerprint that distinguishes my voice from a piano.

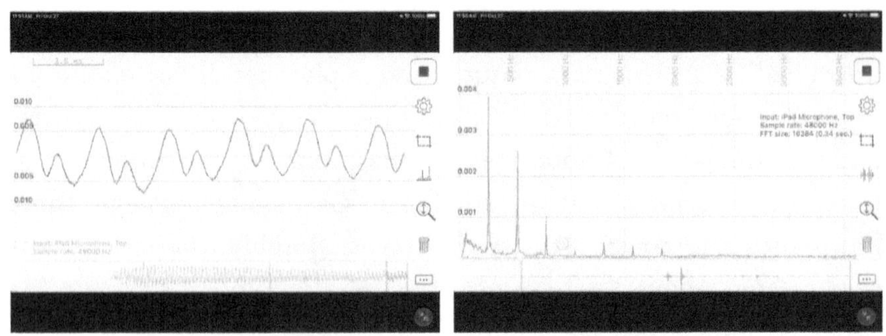

Figure 9.4. Figures 9.4a and 9.4b show the waveform and spectral graph of a C4 key on the piano. Just like my voice, the piano tone includes multiple frequency components, but none stronger than 261.6 Hz.

The spectral signature, in particular, is what we commonly refer to as timbre. This unique sound quality lets us identify different instruments or voices even when they play the same note.

We can at least tentatively conclude that a tone, in the musical sense, generally refers to a single-syllable sound produced by vocal cords or instruments that contains multiple frequency components. Any tone can be thought of as a combination of a few pure tones, each with its own relative strength, as shown in Figures 9.3b and 9.4b.

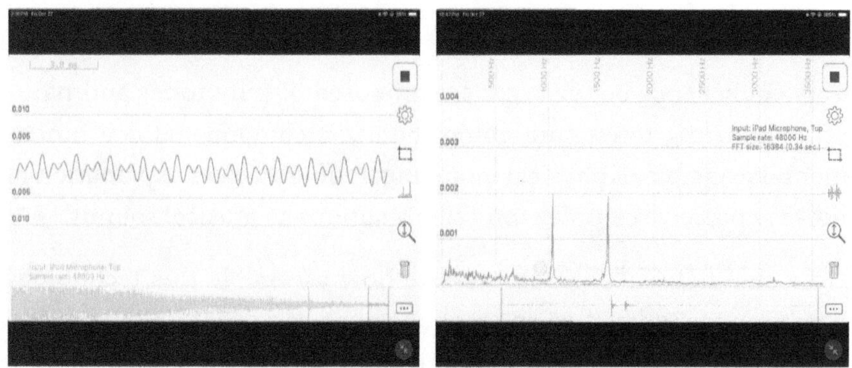

Figure 9.5. The waveform and spectral graph of the C6 and G6 interval played on the piano, constituting a perfect fifth.

It becomes clear that any natural tones or harmonic chords can be reconstructed as long as the spectral graph is known. For example, if I play the notes C6 and G6 together on a piano, forming a perfect fifth interval, their combined waveform and spectral graphs are shown in Figures 9.5a and 9.5b. Alternatively, if I use a computer to generate pure tones at C6 and G6 and play them together, the resulting graphs (Figures 9.6a and 9.6b) look very similar. This supports the idea that a composite tone, made from individual pure tones, can replicate the full sound of a naturally played tone.

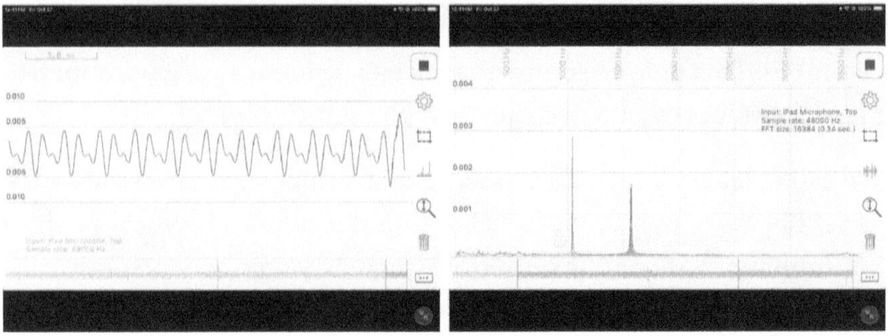

Figure 9.6. The waveform and spectral graphs taken for a perfect fifth interval from computer-generated pure tones of C6 and G6.

This equivalence between tones composed of pure tones and naturally occurring tones containing the same components is a fundamental principle of physical modeling. It offers a clear explanation of how we perceive complex tone combinations as musical sounds.

On a side note, it's also impressive how accessible this technology has become. Today, free smartphone apps can generate pure tones, record audio, and display both waveform and spectral graphs. Back in the 1980s, equipment capable of this analysis cost $30,000 to $40,000 and was made by high-end companies like Keysight (formerly Hewlett-Packard and Agilent) or Rohde & Schwarz. This equipment, including some I helped design at the time, was carefully calibrated against standards set by the National Institute of Standards and Technology (NIST) for accuracy and precision, which explains the high cost. Looking back, it's truly remarkable how far we've come. What once required advanced scientific instruments can now be done with just a phone. Without these affordable modern tools, we would have had to go back to the 1980s to create the kinds of graphs that reveal the core of musical tones in this chapter.

Consonance and Dissonance

According to *Encyclopedia Britannica*, consonance and dissonance describe the emotional responses caused by combinations of tones. Consonance creates a sense of stability and calm, while dissonance produces tension and unrest. These feelings occur when specific tones are played simultaneously.

Although there is no universally accepted definition, consonance is generally seen as a pleasing and stable sound, while dissonance feels harsh or jarring. For example, an octave chord, which is a pair of tones separated by a factor of two in frequency, is usually considered consonant. On the other hand, two neighboring piano keys (a semitone apart) are often regarded as dissonant. Interestingly, this definition suggests that even single-syllable tones (which naturally include harmonics, as discussed earlier) are inherently consonant when produced by a single instrument.

One of the earliest theories explaining consonance and dissonance comes from Pythagoras, who suggested that tones produced by strings with simpler ratios, such as 2:1 or 3:2, are considered more consonant. His observations showed that, even without understanding the reasoning, listeners experienced greater sonority with simpler ratios. Aristotle later acknowledged the existence of both pleasant and unpleasant intervals. However, he emphasized the subjective aspect of musical perception, asserting that the ear itself is the final judge of whether something is consonant or dissonant. Although they used different approaches, both emphasized a fundamental truth: our perception of consonance combines objective acoustic properties with subjective experience.

The following sections aim to bridge that gap by exploring how physical approaches link physical acoustics with psychological perception, giving equal importance to both microscopic neural activity and our large-scale emotional responses.

Auditory Neural Impulses in Short

Understanding how the brain perceives consonance starts with a brief overview of how neural impulses are generated in the inner ear. As explained in Chapter 4, any sound, including musical tones composed of multiple frequencies, is processed by the cochlea in the inner ear.

The neural impulses originate from the organ of Corti, which contains approximately 20,000 hair cells linked via the basilar membrane. When sound waves cause these hair cells to move in synch with the waveform, they release chemicals that generate electrical charges. If the charge surpasses a certain level, it sends an electric impulse, or neural spike. These impulses are then quickly transmitted through auditory neurons, forming the basis for perception.

Once generated, these impulses follow a clearly defined neural pathway (Chapter 6). They first pass through the brainstem nucleus, then move toward the auditory cortex, where sounds are interpreted for meaning and emotion. The patterns of impulse firing influence how we ultimately perceive consonance and dissonance.

The regularity, timing, and strength of these impulses are essential to the physical models discussed in the following sections.

Early Scientific Explorations

The study of consonance and dissonance, especially when two tones are played together, has been explored for thousands of years. Historically, the main explanation came from Pythagoras around 500 BCE. He discovered that simpler frequency ratios between two tones produce more consonant sounds. For example, an octave, which has a 2:1 frequency ratio, sounds very consonant. Conversely, a semitone, with a more complex 15:16 ratio, is generally dissonant.

Pythagoras performed experiments with strings of different lengths and discovered that, when other factors like material, tension, and thickness stayed the same, frequency was inversely proportional to the string length. This insight allowed Pythagoras to modify the pitch of the sound he produced and provided a method for studying chords. For instance, two strings where one was shortened to two-thirds the length of the other produced the second-most consonant interval after the octave. This interval is now called the perfect fifth in Western music.

In 1877, Hermann von Helmholtz proposed a groundbreaking theory. In his "beat theory," he suggested that consonance and dissonance result from how complex tones (with many harmonics) interact. When two tones have simple harmonic relationships, their waveforms align well. However, when the frequencies are more complex, the harmonics don't align properly, causing beating—a type of interference that sounds rough or unpleasant. According to Helmholtz, dissonance increases with the number of mismatched harmonic pairs. His theory clearly explains why some tone combinations feel stable while others sound jarring.

However, Helmholtz's time predates modern neuroscience and neuroimaging. His theory didn't consider the brain's role in shaping our perception of consonance. Specifically, it ignored that dissonance can occur without harmonics and that frequency differences can influence consonance in surprising ways. Some intervals may even seem more consonant as their frequency separation increases, which conflicts with the idea that beating alone explains everything.

Helmholtz's theory also doesn't hold up well when tones are played sequentially instead of simultaneously. However, experiments show that sequential intervals with simple frequency ratios still sound more consonant, another sign that more factors are involved than just harmonic beating.

Neurological Responses to Consonance

Modern neuroscience offers deeper insights. One effective method involves comparing brain responses to musical stimuli in healthy people versus those with lesions in specific auditory regions. Studies have shown that people with damage to the auditory cortex struggle to distinguish between consonant and dissonant tones. This challenges Helmholtz's idea that musical perception is governed solely by the inner ear's peripheral mechanisms. Instead, it suggests that higher-level auditory processing—especially in the brain's cortical pathways—plays a vital role.

Technologies like EEG and fMRI now enable the tracking of brain areas that respond to specific stimuli, offering precise temporal and spatial resolution. When people listen to pairs of pure tones, results show that consonance perception heavily relies on how the brain processes pitch relationships in the cerebral cortex. In short, consonance is not just the absence of roughness; it's a cognitive phenomenon, shaped by how the brain interprets pitch combinations.

These conclusions are based on analyzing auditory evoked potentials, which measure electrical activity in the brain. For example, responses were strongest for the perfect fifth (7 semitones) among intervals such as 1, 4, 6, 7, and 9 semitones. This shows that the brain reacts more strongly to intervals with simple frequency ratios, even without harmonics. Obviously, these neurological determinations of consonance are consistent with the massive subjective study mentioned earlier in this chapter and shown in Figure 9.1.

This electrophysiological evidence strongly supports behavioral findings that we naturally prefer such intervals, reinforcing the idea that the brain's structure and function are tuned to favor simple, consonant relationships in sound.

Physical Scientific Approaches

This section presents two scientifically supported yet physically different models that examine how we perceive consonance in music. Both models rely on plausible assumptions and rigorous calculations to reach a common conclusion: simpler frequency ratios tend to create stronger consonance. As we will see later, real-world perception may involve elements from both models.

<u>Common Setup</u>

Both models start with the same initial point: two pure tones, PT1 and PT2, enter the inner ears, where they interact with the cochlea. Within the cochlea, sensory hair cells vibrate in sync with the incoming waveforms—each aligning with the tones' specific temporal patterns and frequencies, as illustrated in the waveform graphs in Figures 9.7a and 9.7b.

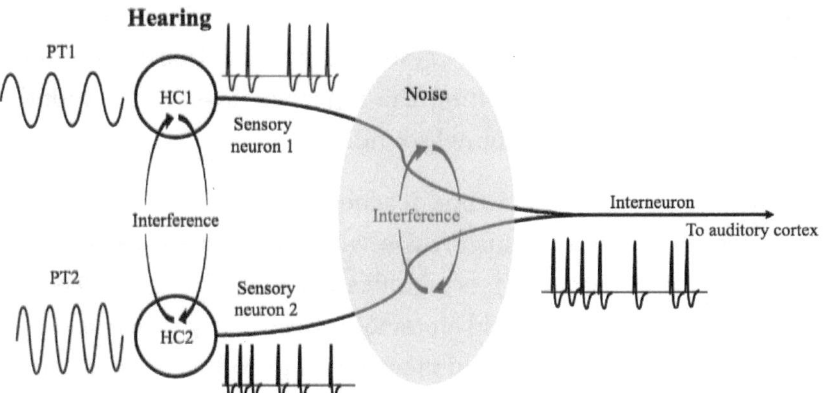

Figure 9.7a. The common setup for the two modeling consonance calculations. This figure shows how the pure tone waveforms in the time domain are converted into electrical impulses. Notice that the conversion does not retain the pure tone waveforms.

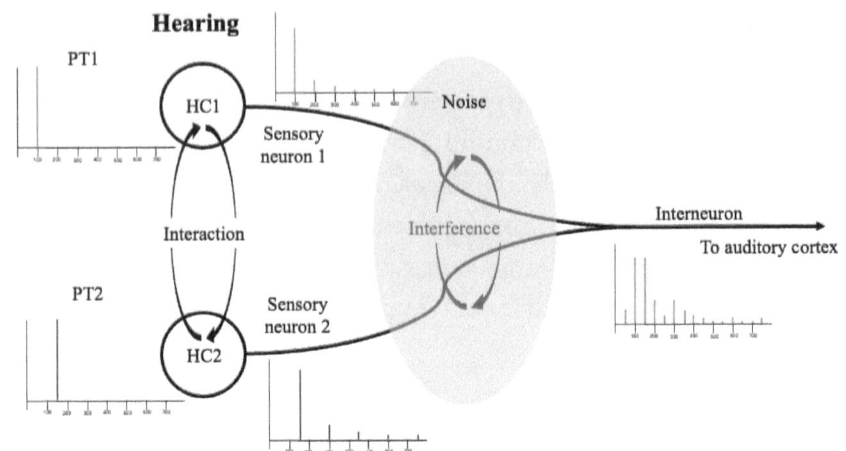

Figure 9.7b. The common setup for the two modeling consonance calculations. This figure illustrates how the pure tone spectral graphs are altered due to the conversion.

These waveforms are then converted into neural impulses, as previously described. The pure tones activate the hair cells, which convert physical vibrations into electrical signals, or spikes, that travel through the auditory nervous system. However, this conversion is inherently non-linear, meaning the inner ear does not respond in direct proportion to the original sound.

When air pressure from a tone reaches the eardrum, the resulting impulses do not exactly mirror the waveform. As shown in Figure 9.7a, the output signal differs significantly from the input. The hair cells function like threshold detectors, firing only when a certain level is reached. This threshold mechanism introduces faster components than those in the original waveform. As a result, the spectral graph of the impulse train, Figure 9.7b, shows many high-frequency harmonics that weren't present in the original pure tone.

MODEL 1: PHASE LOCKING BY CROSSTALK

The first model relies on non-linear interactions between the two pure tones. These interactions tighten the conditions for generating impulses and create what's been described as a "dragging" effect. In this model, hair cell HC1 is stimulated by tone PT1 and produces strong, synchronized neural impulses—about 0.100 volts (Chapter 5) in amplitude. Some of this electrical signal can "leak" and be detected by neighboring hair cells, such as HC2, which is tuned to PT2.

To appreciate the importance of this interaction, consider this: the signal strength received by a cell phone is often just 0.000007 volts, yet it can still produce clear conversations. That's only a ten-thousandth of the strength of a neural impulse. Therefore, it is not a stretch that nearby hair cells detect each other's signals, effectively creating interference or what we might call biological crosstalk.

To understand how this results in phase locking, picture two backyard swings of different lengths. Swings of varying lengths naturally move at different speeds. Now, imagine two people riding on the swings. Sometimes, the swings come close enough for the riders to "nudge" each other gently, causing their movements to adjust subtly. Over time, the swings may begin to synchronize, even if they began out of sync.

In this analogy:

- PT1 and PT2 are the two swings.
- The nudge is the small electrical interference between neighboring hair cells.
- Phase locking occurs when the swings (tones) move in sync.

When the two swings (PT1 and PT2) are nearly in sync — meaning their phases are close but not identical — even a small nudge can cause them to lock into the same phase. In physics, phase refers to the position within a cyclic process, often measured from 0° to 360° for a full cycle. When the phases align, the two tones are said to be locked, and both are dragged and forced to slightly alter their frequencies to stay synchronized.

While the original tones (PT1 ~ PT2) may need only slight adjustments, the adjusted ones remain fixed and consistent (PT1 = PT2). As the ratio PT2/PT1 decreases from 1 (indicating the tones are moving apart), this phase-locking effect weakens. The two tones start to act more independently, like two swings that cross paths only occasionally.

Figure 9.8 displays a series of phase-locked states, represented by horizontal bars. These show frequency ratios where the tones synchronize and sound consonant. The strongest phase-locked ratios include 2/3 and 1/2, which correspond to the perfect fifth and the octave, respectively.

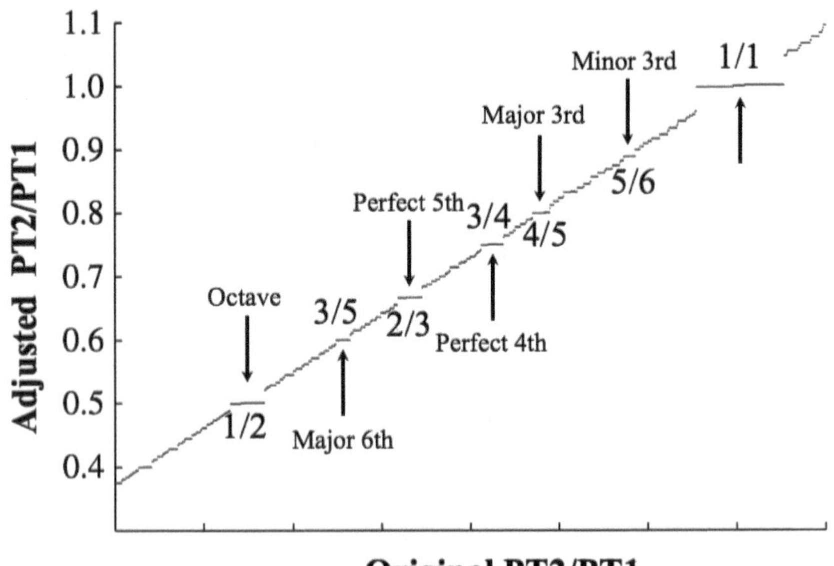

Figure 9.8. This chart shows a series of phase-locked states, marked by horizontal bars. These represent frequency ratios where the phases of the tones synchronize and thus sound consonant. Among the strongest phase-locked ratios are 2/3 and 1/2, which correspond to the perfect fifth and the octave, respectively.

To clarify the swing analogy:

- At a 1/1 ratio, the swings move together perfectly at all times.
- At a 2/3 ratio, one swing completes two cycles for every three of the other. They periodically align and lock in sync for a moment before falling out again, then realign as the cycles repeat.
- Other ratios like 3/4, 3/5, and 5/6 (corresponding to the perfect fourth, major sixth, and minor third) also produce weaker but noticeable phase-locking states. These were called "just" intonations by the Ancient Greeks and are also perceived as consonant.

These key ratios also match numbers in the Farey sequence of the sixth order, forming a visual pattern known as the devil's staircase—a stepped curve showing transitions between locked states (more on this later).

<u>From Physical Interference to Perceptual Consonance</u>

Returning to the swing analogy, imagine that as PT2/PT1 approaches the 2/3 ratio, PT2 is more easily pulled into phase by PT1, or vice versa. The closer their frequencies are to a "just" ratio, the easier it becomes to achieve phase lock.

The length of each bar in Figure 9.8 shows the strength and duration of that phase-locking interaction. Longer bars indicate the two tones have their phases in sync with wider original frequency differences. In terms of hair cell behavior, this means that HC1 and HC2 fire at regular intervals over a broader range of PT1/PT2 values, improving the sense of consonance.

When tones phase-lock, they "swing" together, at least briefly. Because these phase-locked states align closely with our subjective experiences of pleasantness, it's reasonable to conclude that this connection is not coincidental. Instead, phase locking and synchronization may offer a strong scientific explanation for why we perceive certain intervals as consonant.

This model starts with first-principle physics and considers only minor, measurable interactions between the tones. Still, it produces results that align well with subjective musical experience: consonance occurs at frequency ratios that cause phase-locked synchronization in the auditory system.

The appearance of the devil's staircase, a mathematical construct derived from the Farey sequence, adds another layer of intrigue. Although Farey sequences are purely mathematical and have no inherent connection to music or psychology, their presence here suggests that consonance and dissonance are rooted in mathematics and physics, not just subjective emotion. Of course, biology also plays a role in this, as PT1 and PT2 can sense each other, and their neural impulses interact.

In other words, our musical instincts might stem not from cultural influences, but from the fundamental properties of cyclical systems, neural interference, and frequency ratios, which are as objective as the math that describes them.

Model 2: The Noisy Brain

The second approach introduces a realistic, noisy environment into the neural impulses experienced in the brain. Referring to Figure 9.7, one can imagine that each location—such as HC1, HC2, the sensory neurons, and the interneurons—is immersed in noisy molasses. Additionally, assuming the noises experienced by HC1 and HC2 are unrelated, any effect on perceptions will come solely from independent noises and how the impulses from HC1 and HC2 are combined. The neural impulses and noise generated by HC1 felt by HC2, and vice versa, can be seen as disturbances with varying strengths relative to each other, as discussed regarding the lengths of those horizontal bars in Figure 9.8. Noise can also appear as phases; in other words, the uncertain phase would influence when HC1 and HC2 fire, creating an equivalent phase nudge between HC1 and HC2, as discussed in Model 1, but in this case, based on probability.

The follow-up calculation determines the probabilities, or the degree of regularity, of nerve impulse firing over time as PT2/PT1 increases. The regularity is assessed by the probabilities of the time intervals between firings, as shown in Figure 9.9. This pattern of impulse firing assumes no prior knowledge of possible consonance or musical psychology. Note that the time scale is relative, with 10 in Figure 9.9a (upper left) referring to an audio cycle of PT2's C4, which is around 261.6Hz. In the case of the 1/2 octave in Figure 9.9a, PT1 is an octave higher (C5) and requires half the time to fire the next neural impulses.

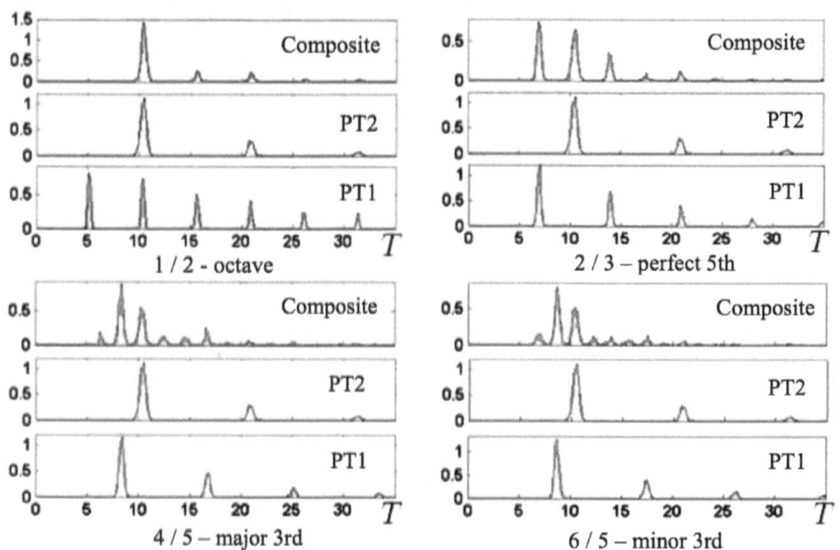

Figure 9.9. *The regularity of neuron firing is measured by the probability of the time elapsed between firings. The peaks indicate the time between firings, and the spacings are equivalent to the firing frequency.*

The composite tone, displayed as the upper traces in Figure 9.9 at different ratios, shows peak probabilities at intervals of one audio cycle, half a cycle, one and a half cycles, and so on. These probability peaks suggest that firing occurs regularly, i.e., with high consistency.

It is also observed that the perfect fifth (2/3) in figure 9.10b (upper right), the major third (4/5) in figure 9.9c (lower left), and the minor third (5/6) in figure 9.9d (lower right) intervals show regular firings with varying levels of "regularity." They also happen to be the just intonations mentioned earlier by the Greeks.

Model 2 functions as a purely mathematical and physical model of how hearing transforms into perception without interaction between the pure tones. Still, it aligns with the simple interacting pure tones and phase-locked conditions in Model 1, from which consonant conditions directly emerge from the calculations.

A Unified Interpretation

Although rooted in different principles—one deterministic and the other probabilistic—both approaches arrive at the same conclusion: simple frequency ratios lead to increased neural firing regularity and thus consonance.

- In the deterministic approach, phase locking causes two pure tones to merge and reinforce each other, resulting in regularly fired impulses.
- In the probabilistic approach, noise-affected firings still follow predictable patterns when frequency ratios are simple.

It is therefore reasonable to equate deterministic synchronization with the firing regularity observed in the probabilistic model. Together, these perspectives provide a unified physical explanation for how we perceive tonal quality. Neither model needs any input from human psychology. Yet, the physiological processes they describe directly lead to the subjective feeling of consonance. What we see as pleasant or unpleasant isn't random; it's the result of computation based on physics and math.

Pythagoras, who took a scientific approach to consonance, focused on objective frequency ratios. Aristotle, on the other hand, viewed consonance as a subjective experience, resulting from how sound is processed in the "ear" and the mind. This chapter on psychoacoustics shows that both perspectives are valid and complement each other.

In truth, the difference between objective hearing and subjective perception might be more philosophical than practical. The broad understanding of sound and the detailed physiology of hearing are two aspects of the same psychoacoustic concept.

Empirical Correlation With Human Judgement

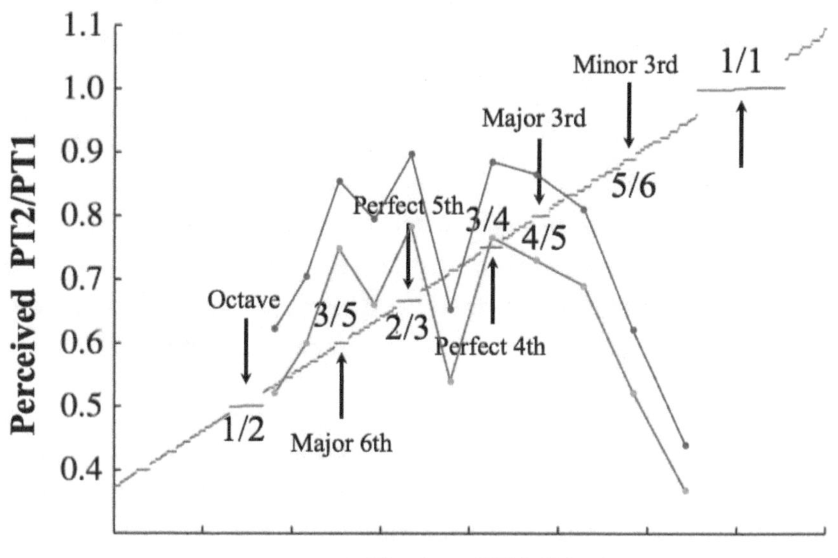

Figure 9.10. The overlay of Figures 9.1 and 9.8 is plotted against PT2/PT1 ratios. The peaks in pleasantness align closely with simple frequency ratios, directly connecting the empirical human judgments with the physical phase-locking model and noisy auditory neural environments.

In Figure 9.1, we demonstrated how the pleasantness of tone combinations—as rated by a large number of untrained participants—varies with interval size. In Figure 9.10, we overlay Figures 9.1 and 9.8 using a recalibrated scale aligned with the original PT2/PT1 ratios. The result is striking: the peaks in pleasantness align closely with simple frequency ratios, directly linking human judgments with the physical phase-locking model and noisy auditory environment.

Discussion

The history of structured music is relatively brief, likely less than 10,000 years old. Over time, humans invented chords and harmonies to enrich musical expression, which has an even shorter history. What these inventors didn't realize is that their creations to please our hearing are also rooted in natural and scientific principles. Humans are naturally attracted to consonant sound combinations, not only for cultural reasons but because such combinations match our natural and biological preferences.

Western culture, in particular, celebrates the invention of harmonies, tracing them back to Pythagoras in ancient Greece. But as this chapter suggests, naturally occurring tones are inherently consonant, and the "invention" of harmony is more like a rediscovery of nature's patterns. This observation doesn't diminish the value of harmony in Eastern traditions, such as a cappella, stone, and metal ensembles in ancient China.

Since the 15th century, harmony has been developed and integrated into Western music, forming the foundation for the complex contrapuntal structures of Johann Sebastian Bach, especially in his *Well-Tempered Clavier*. The fugue, for example, is a form that relies on multiple melodies overlapping and intertwining. These melodies are arranged to produce consonant intervals at regular points, appealing to our subconscious physiological preferences, even if composers aren't always consciously aware of it.

A simple example is the French nursery rhyme "Frère Jacques," as shown in Figure 9.11. When the song is sung in a round (canon), the overlapping parts form C–E (major third) and E–G (perfect third) intervals, both consonant and pleasing to the ear. Incorporating regular, pleasant consonant intervals throughout makes the song overall engaging and satisfying.

Frère Jacques

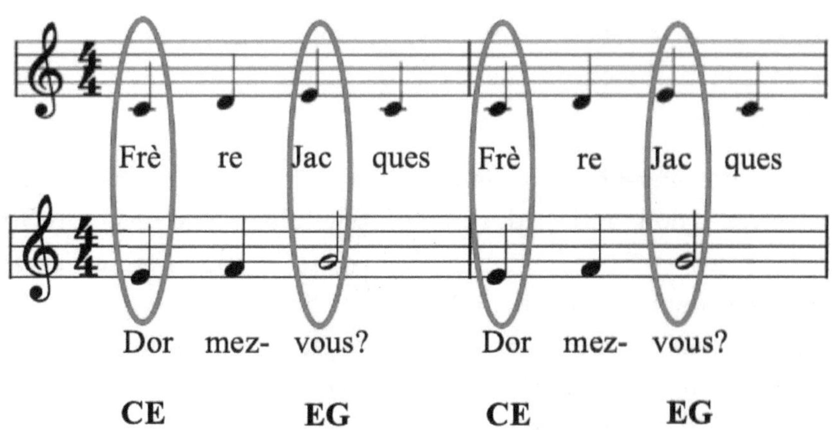

Figure 9.11. *When the nursery rhyme "Frère Jacques" is sung in a round (canon), the overlapping parts form C–E (major third) and E–G (perfect third) intervals, both consonant and pleasant to the ear. The consonant intervals in this example happen regularly throughout the round. Interspersing pleasant consonant intervals throughout makes the song pleasant and satisfying.*

It's important to recognize that dissonance is not inherently "bad" or undesirable. Even the "Devil's chords" serve specific purposes. It plays a crucial role in musical expression. In jazz and popular harmony, chords are often labeled by root and modified with symbols that indicate different levels of tension. Not every chord needs to be perfectly consonant. Musical styles from Baroque to Romantic to modern jazz frequently use "tensions," notes that create dissonance relative to the bass.

These tensions generate contrast and suspense, which are then resolved by returning to consonant chords. This interaction between dissonance and resolution is key to musical storytelling. In this context, counterpoint—the interaction of independent melodies—is sometimes viewed as separate from harmony but also serves as a way to balance tension and relaxation.

Summary

This chapter examines several scientific interpretations of how our senses perceive tones. Specifically, humans judge tones as pleasant or consonant based on the intervals played at specific frequency ratios.

Before the invention of musical instruments, humans always perceived good and bad through physiology and evolutionary instinct. The most common characteristic of comfortable human sounds is that they contain multiple audio frequency components. In terms of hearing mechanics, human tones resemble multiple pure tones reaching the cochlea at the same time. This understanding forms the basic assumption of the scientific study of how humans perceive harmony, including consonance and dissonance.

This chapter also aims to simplify the human hearing-perception system for single-syllable tones into a manageable problem and offers a fundamental interpretation. Despite the oversimplification, we have gained some scientific clarity about tonal perception. Among various theories about our innate preference for harmony with simple ratioed tones, this chapter's physical models and evidence suggest a physical and biological basis for our perception of consonance and dissonance. These preferences are not merely cultural; they are rooted in the structure and function of physics and the human nervous system. They also align with findings from neuroscientific studies and offer a new perspective on how cross-disciplinary research can help explain complex phenomena, such as our cognitive reactions to harmony and music.

Bibliography

1. "Individual Differences Reveal the Basis of Consonance," Josh H. McDermott, Andriana J. Lehr, and Andrew J. Oxenham, *Current Biology*, **20**, 1035-1041 (2010).
2. "Perception of musical consonance and dissonance: an outcome of neural synchronization," I. S. Lots and L. Stone, *J. R. Soc., Interface*, **5**, 1429 (2008).
3. "Impulse-coupled neuron models as investigative tools for musical consonance," B. Heffernan and A. Longtin, *J. Neurosci. Methods*, **183**, 95 (2009).
4. "Regularity of Spike Trains and Harmony Perception in a Model of the Auditory System," Yu. V. Ushakov, A. A. Dubkov, B. Spagnolo, *Physical Review Letters*, **107**, 10813 (2010).
5. "Spike train statistics for consonant and dissonant musical accords in a simple auditory sensory model," Y. Ushakov, A. Dubkov, B. Bernardo Spagnolo, *Physical Review E*, **81**, 041911 (2010).
6. "On the sensations of tone as a physiological basis for the theory of music," H. Helmholtz, *Dover Publications, New York, NY* (1877).
7. "Children's discrimination of melodic intervals," Schellenberg, E. G. & Trehub, S. E., *Dev. Psychol.* **32**, 1039–1050 (1996). doi:10.1037/0012-1649.32.6.1039
8. "Cortical deafness to dissonance," Peretz, I., Blood, A. J., Penhune, V. & Zatorre, R. 2001 *Brain* **124**, 928–940 (2001).
9. "Neurobiological foundations for the theory of harmony in Western tonal music," Tramo, M. J., Cariani, P. A., Delgutte, B. & Braida, L. D. *Ann. NY Acad. Sci.* **30**, 92–116 (2001).

10. "Cortical processing of musical consonance: an evoked potential study," Itoh, K., Suwazono, S. & Nakada, N. 2003. *Neuroreport,* **14**, 2303 –2306 (2003).

11. "Mental rotation in visual and musical space: Comparing pattern recognition in different modalities," Korsakova-Kreyn, M. and Dowling, W. J., *Univ. Texas, Dallas Publication* (2008).

12. "Bach, Escher, and Mental Rotation: An Empirical Study in the Perception of Visual and Melodic Congruency," Korsakova-Kreyn, M. *IAEA Conference,* (2014).

13. "Musicophilia, Tales of Music and the Brain," Oliver Sacks, *Alfred A. Knopf Publisher,* (2007).

14. "The Descent of Man", Charles Darwin (1871).

15. "The Prehistory of Music: Human Evolution, Archaeology, and the Origin of Musicality," Ian Morley, *OUP Oxford Publisher* (2013).

16. "100 Metronomes Ticking at Different Speeds Fully Synchronize While Sitting on a Floating Table", laughingsquid.com/100-metronomes-synchronize/

17. "Timbral effects on consonance disentangle psychoacoustic mechanisms and suggest perceptual origins for musical scales," Raja Marjieh, Peter M. C. Harrison, Harin Lee, Fotini Deligiannaki, Nori Jacoby, *Nature Communications,* **15**, 1482 (2024). DOI: 10.1038/s41467-024-45812-z.

5. Cognition and reward

Each phase involves distinct neural activities and draws on different scientific disciplines, languages, and methodologies. This division is based on phenomenology, serving as a way to describe the observable steps in how music affects our brains.

A guiding principle runs through all these phases: the theory of statistical learning and predictive coding (SLPC), first introduced about 40 years ago. SLPC proposes that our brain continually learns from patterns and predicts what will happen next. This theory forms the basis of much of how musical signals are processed in the brain. However, SLPC remains a descriptive, or phenomenological, model.

To ground SLPC in something more fundamental, we turn to physics, specifically, the second law of thermodynamics (SLTD). This law, which has stood unchallenged and unviolated for centuries, states that systems naturally evolve toward states of lowest energy. In the context of music, SLPC operates under the mandate of SLTD: the brain strives for efficiency, predicting and adapting to minimize the energy expended for the unexpected.

We borrow from engineering to visualize the details of SLPC: the control system model. Feedback and feedforward loops in such systems provide an analogy for signals flowing in both upstream and downstream in our brain's organized hierarchy. Chapter 4 introduces this concept with familiar everyday examples, helping us see how these principles apply to music processing.

This approach is not meant to oversimplify. While our control-system analogy might seem reductionist, the underlying science, especially at the microscopic level, remains the main thrust of the understanding of SLPC. Later chapters (6, 7, and 8) explore the neuroscientific details more deeply.

Combining SLPC and SLTD with neuroscience provides a solid framework for understanding music cognition. This guiding principle helps us put the pieces together into a coherent whole, revealing the pathways through which musical messages move within the brain.

Of course, neuroscience alone cannot solve all the puzzles. Since SLTD forms the foundation of SLPC, physics plays a crucial role. Add in the contributions of biology and chemistry, and the study of music becomes a truly interdisciplinary effort. For example, Chapter 9 explores consonance and dissonance, qualities that physicists have studied extensively, from a physics perspective, providing insights where neuroscience alone has not. Only through such cross-disciplinary efforts can we get closer to understanding the science of musicality.

The previous pages summarize this book, which attempts to shed intuitive light on the reasons behind our obsession with music, through the inherent musicality in each person and group. But does it answer all the questions asked in Chapter 1? Do we finally have an answer to the "big" question: why are humans so passionately drawn to music? Let's revisit these questions in the following using what we have learned. I am sure you can make a better-informed judgment for yourself before closing the book.

Have We Answered the Questions Posed in Chapter 1?

What is Music?

One of the most basic questions, "What is music?", should now feel more straightforward. By studying how the brain processes music, we've discovered that our auditory system breaks music into parts that match what we naturally think of as its fundamental elements. Once music enters the auditory sensory system, it is generally divided into two main categories: spectral (related to pitch and tone) and temporal (related to timing). At this initial stage, music can be described as tones arranged in time variation and in spectral components.

Based on this auditory analysis of the music, the brain constructs meaning and assigns emotional value through pathways in the auditory ventral and dorsal streams. Here, music becomes more than just organized sound; our brain has decoded the melodic and rhythmic patterns that hold intellectual significance. These qualities serve as an additional layer to music's definition.

Finally, the reward system steps in. It processes these musical inputs, interprets their meaning, and translates them into feelings and physical responses, such as being sad or happy, tapping your foot or swaying to the beat. This forms the highest level of the definition: music as an experience that stirs emotion and movement.

In summary, music is an organized temporal arrangement of tones with a layered structure. The first layer consists of pitches and timing. The second layer involves encoded meaning, as well as patterns of tone and rhythm that convey ideas. The third layer is an implicit "action plan," instructing us to respond with emotion and movement according to that plan.

This is a functional definition of music based on a layered structure, which isn't the only way to think about music, but it reflects how the brain processes it based on this book's narration.

Conferring Evolution Advantages?

Another long-debated question is whether music provided humans with an evolutionary edge. After exploring topics related to musicality, we can offer several observations.

1. The brain regions responsible for music perception, cognition, and reward seem to be unique to humans and lack direct equivalents in other animals. These areas are

located in the most recently evolved parts of the mammalian cortex, and music mainly activates them and hooks back to our primal responses. Has this ability been involved with human evolution toward better or worse, adapted to survival, and our species' benefits?

2. Musical functions rely on neural systems also used for language, raising the classic question: which came first, music or language? This remains one of the most significant scientific controversies.

3. Musical synchronization, characterized by moving in sync to a beat, can boost group cohesion and may give some communities a survival edge. However, claiming music as an evolutionary advantage is a stretch since structured music has a relatively short history, considering the time it takes to see any evolutionary outcome of any traits.

4. Music activates some brain areas that other traits, like language, movement, and emotion, rarely engage at the same time. But musical activities themselves are not inheritable, thus have no impact of genetic predisposition in the evolutionary sense.

5. Despite extensive research, no "music gene" has been identified, unlike the FOX2P gene for language.

6. Following the suggestions by Smith, Hume, and Schopenhauer (Chapter 1) that music's collective activities, seeking synchrony with fellow humans, may have contributed to our early evolution once we got ourselves into music. One thing is certain, though; musicality indeed drives the motivation for synchrony (Chapters 7 and 8) in group settings. This group dynamics is thought to bring about empathy and, in turn, encourage sympathy, morality, ethics, and justice, all of which enhance group cohesion and could confer evolutionary advantages.

Taken together, these points still do not lead to a satisfactory conclusion regarding music's role in early human evolution. It remains unclear whether it is likely to do so in the future.

However, the existence of neural plasticity—the brain's ability to rewire itself—means that intensive music training can reshape brain structure. Some limited evidence hints that this plasticity could be passed on epigenetically to future generations, potentially predisposing them to nurture existing musicality better. Still, neural plasticity simply allows for the "moldability" of musical ability and should not be considered a musical genetic trait.

Why Do Musical Tastes Change?

We also asked why musical tastes change over time. The explanation involves statistical learning and predictive coding (SLPC), guided by the principles of the second law of thermodynamics (SLTD). The music we grew up with becomes ingrained in memory (SL), influencing our preferences. These preferences interact with the brain's reward system and generate feelings of (un)familiarity and enjoyment per the predictive coding (PC). As cultures and societies develop, so do music and our tastes. This shift in taste is not random but a natural result of cultural changes combined with the brain's learning and predictive mechanisms.

Why Aren't We All Musical Geniuses?

Another question close to my heart: if musicality is innate, why am I not a Beethoven or a Mozart? Why are some people prodigies while others are not? The answer mirrors language: just as few become Shakespeare, and there is only one Bach. Musicality provides the foundation, but genius develops through extensive exposure and training, taking advantage of the SLPC process. I lack both, so it's no surprise I'm not a master composer or performer. There are, of course, individuals who are exceptionally talented in the art of literature or music, which sets them apart from the rest of us.

Music And The Brain: a Broader Impact?

Our elaboration through the chapters shows that music activates more parts of the brain than most other activities, including language. As explained in Chapters 6, 7, and 8, music triggers networks involved in cognition, emotion, and psychology, which improves connectivity and strengthens neural pathways. This enhancement makes the Mozart effect, or improved cognitive function through music exposure, plausible. Neural plasticity ensures that many of these benefits last throughout lives, allowing them to be reinforced later.

Have We Answered the Big Question?

So, have we answered why we have an obsession with music? In a small way, the very act of writing this book and your act of reading it demonstrate that we all have a desire to learn more about music. On a larger scale, we share this obsession as described extensively in Chapter 2 because we are wired for musicality. Beyond Chapter 2, this book elaborates on the common ability to create and interpret music with pitch, melody, structure, and emotion that can only be explained with something fundamental built into our biology and the underlying physics. The clear answer to that question is "musicality," and our obsession seems a natural consequence.

Eventually, every question circles back to the reward system. It is this system that drives us to seek music repeatedly. But the very possibility of this reward loop relies on something deeper: our innate musicality. Without it, the entire process of perception, cognition, and pleasure would fall apart.

What Comes Next?

Now that we've addressed the fundamental question of the cause of obsession, although still leaving much to answer, what could be the future thrusts for neuroscience and the science of musicality?

Music engages more regions of the brain than nearly any other human activity, including language. Our ability to make music likely predates language, dating back to the earliest days of humanity. Unlike language, which offers obvious survival advantages like coordinating hunts or farming, music provides something less tangible but arguably more meaningful: it stimulates intricate, rich networks in the brain that influence our emotional and social lives.

Music's complexity arises from the intricate interaction of melodies, harmonies, rhythms, structures, and lyrics. It allows us to express emotions that words often cannot capture. Some musical experiences are so layered and subtle that our current languages can't fully describe them; maybe we would need an entirely new language for that. Music acts as a direct path to these deep, often indescribable feelings. Gaining a deeper understanding of the neuroscience of music could offer more detailed insights into these processes.

We have tried to answer the compelling question of why we are so enamored with music. But the quest offers valuable insights into how the brain processes sound, rhythm, and harmony, uncovering the mechanisms behind learning, memory, emotion, and sensory integration. In this way, music becomes a powerful scientific tool for exploring the brain's most complex functions, which help define our humanity.

The future of musical neuroscience also relies on cross-disciplinary collaboration, as demonstrated in Chapter 9. For example, physics has provided insights into musical phenomena that neuroscience alone has not solved. My background in physics and engineering has influenced how I approach these questions, allowing me to use models and principles to simplify complexity and uncover intricate mechanisms. By integrating perspectives from neuroscience, physics, psychology, and engineering, we can strive for a much more comprehensive understanding of music than ever before.

So far, we have identified general brain regions activated by specific elements of music. However, their precise locations and boundaries remain unclear, and they often differ between individuals. One major question interests me: How do neurons cluster to process individual and entire units of music? This is still largely unknown, and much research remains to be done.

Currently, we divide music into familiar elements like melody, harmony, and rhythm, and link them to different areas of the brain. For instance, melodies and rhythms are processed in separate regions; movements and emotions are also involved in distinct regions. But is this the best way to analyze music? If we approached it differently, could we find a clearer, simpler explanation for how the brain understands music? Does the aforementioned simpler correspondence (Chapter 7) between elements and brain regions/functions exist? Or is the overlapping and repetitive nature of the functions and regions the norm for all neural processes? Neuroscience has clarified similar questions about language, finding that the one-to-one mapping does not hold true. But how about music? This could be interesting to explore.

Science of Music and Art of Music

For centuries, some have feared that scientific analysis might strip music of its mystery, the same mystery that once infused religious rituals, inspired great composers, and defined cultural identity. Does dissecting music into "dry facts" diminish its beauty? I believe this fear is based on a false premise. Science and art are not adversaries; they are allies. Understanding how music works only enhances its wonder through its emotional impact.

I'm not a musician, but I've always enjoyed music from different cultures—anything that pleases my ear. For a long time, only a few pieces, like Tchaikovsky's 1812 Overture, gave me chills. However, something changed during the nearly two years I spent researching and writing this book. At first, I believed musicality was an either/or quality; you had it, or you didn't. Then, as I delved into the science of music—how it's processed, how it affects brain activity, and how it affects my emotions—I noticed something surprising: I was becoming more attuned to musical nuances. Many of the music pieces I had known for years now gave me goosebumps. Music started to move me in new and profound ways. This experience taught me something profound: exploring music scientifically does not make it colder or more clinical. On the contrary, it nurtures musicality. It makes us more emotionally engaged. If scientific exploration could heighten my emotional response as a non-musician, imagine what it could do for professional musicians.

The more we learn about how the brain responds to music, the more inspiring it becomes for composers, performers, and listeners alike. Whether or not someone understands the neuroscience behind the music, its creators will still craft melodies that express their deepest emotions, and listeners will still feel joy, chills, and a sense of connection. Dancers will continue to move instinctively to rhythms that touch the soul. But knowing that music is broken into parts in our brain to be analyzed and then reassembled for reward and pleasure adds even more meaning to the music we cherish.

By asking a simple question, "What is the root cause of our obsession with music?" this book has led us to an insightful discovery: our fascination with music comes from an inherent human ability to perceive, understand, and create music. In other words, it springs from our innate musicality.

Throughout this journey, we've thoroughly explored musicality and built a foundational understanding of how it arises from the basic principles of physics, biology, and neuroscience. This exploration not only broadens our knowledge of music but also strengthens our emotional connection to it, whether in its creation or the pleasure of listening.

Music inspiration and knowledge are not opposing forces; they reinforce each other precisely because musicality is hardwired into us.

Bibliography

1. "Is Music an Evolutionary Adaptation?" Levitin, D. J., *The Cognitive Neuroscience of Music*, 123-145 (2014).
2. "Behavioral epigenetics: How nurture shapes nature". Powledge T, *BioScience*, **61** (8), 588–592 (2011). doi:10.1525/bio.2011.61.8.4
3. "Is Music an Evolutionary Adaptation?" David Huron, *Annals of the New York Academy of Sciences* **930** (1), 43 – 61 (2001).

About the Author

Chris Young Kelly is an experimental physicist-turned-engineer whose career has spanned cutting-edge science, advanced technology, and leadership in the high-tech industry. His early research focused on the quantum behavior of materials used in lasers, specifically those that power the nanometer-scale photolithography essential to modern semiconductor manufacturing. He also explored the quantum characteristics of Neptune's and Uranus' atmospheres by recreating their extreme environments under controlled laboratory conditions.

As an engineer, Chris's work has had a direct impact on daily life. He played a role in laying the foundation for today's high-speed communications infrastructure, particularly its optical backbone, which makes possible the massive flow of Internet traffic for streaming media, IoT applications, and data center networks.

Beyond his research and engineering work, Chris rose to become a senior corporate executive in the high-tech industry. He also shared his knowledge in academia by teaching graduate-level engineering courses as a visiting professor at Stanford University. Later, he broadened his impact as an entrepreneur and investor in high-tech startups.

A prolific contributor to the technical literature, Chris has authored over 100 peer-reviewed publications. Today, he applies the same curiosity and analytical rigor to popular science writing, aiming to make complex scientific ideas accessible and engaging for a broader audience.

Outside of his professional life, Chris is an avid reader, traveler, and hiker. He also practices Tai Chi and considers himself a high-handicap golfer. He and his wife, Wendy, live in Northern California with their pets: a tabby cat, Meowmy, and a golden retriever, Buddy.

You can reach Chris at cykelly60@gmail.com.

were more dangerous than back at home.

Ash had just finished speaking to the people of Edihn about the mission he and his father had received from God. He explained how they were free, the importance of looking to the divine for help and how they should never return to the worship of man over God.

The message was an easy one to get across as the red and black kings both fell with without a central figure present during the uprisings. The people of those kingdoms were eager to accept God.

Ash stepped off the stage and walked over to the group of women Ruthie was talking to. She looked up and saw her husband and happily lifted their three month old son over her head so he would take the baby.

He held his son and thanked him as he always did. After the death of the last king, the white king, they wondered why God intervened. The prophecy had no mention of God saving Ruthie. A few month later when they learned they were pregnant. That is when it became clear that when Ash asked God to pass on the prophecy of the forgotten, it was passed on to the child recently conceived between Ash and Ruthie. It was he, who would eventually be born Emmanuel that could not be killed by the hand of man and therefore was saved by God when the king had planned to kill his mother.

"If any of you need someone to talk to please come and find us. We will be in your town until the next full moon." Ruthie concluded her meeting with the women of the town. She had taken her experiences with the king and used them as a building block to help others throughout the world who had trouble coming to terms with their past in the old world of the kings.

She stood and embraced her husband happy to be doing something for humanity. She kissed him and grabbed his hand. She was prepared to drag him back to their tent where

they were staying until they moved on to the next town. Their time in the towns was often spent with others. Their family time was short and rare. Anytime Ruthie saw the opportunity to be away she seized it.

But, before they removed themselves from the town square a young boy came up to them. He kept glancing back at the man and woman standing a few dozen feet behind him, watching him approach his prince and princess.

He looked up at the couple and said in a soft six year old voice, "Excuse me. Grey princess?"

Ruthie kneeled down to talk to the boy at his level.

"Is it true that you can talk to God?" The boy looked at Ruthie which such innocence and interest that it nearly melted her heart.

Ruthie patted the boy on the head, "We can all talk to God. But, Prince Ash here used to hear God's replies."

"Do you hear it anymore?"

Ash looked down at the boy smiling, "No, but Emmanuel hears him now." He holds up his son so the boy can see who he was talking about.

The boy's eyes lit up with excitement. "What does he sound like?"

"Like all the love in the world speaking to your heart in a single moment."

The boy looked down, unimpressed with the description. He looked back at Ruthie, "Princess Ruthie, you are beautiful." He gave the princess a peck on the cheek and ran back to his parents in the square.

The couple chuckled to themselves and walked back to their tent.

Moshe and Mary Maggie

Chapter 10

Conclusion

Recap

A simple question, "Why are we so obsessed with music?" begins this book. We aim to explore and confirm that we are all musical and that our innate musicality lies at the core of this obsession. To do so, we ask a series of critical questions: some are the kinds we casually think about in everyday life, while others probe much deeper. No matter where we start, persistent questioning eventually leads us to science. That raises the central issue: Is our musicality inborn, and how can it be proven scientifically?

To ensure this exploration is meaningful, we establish two compelling facts in Chapters 2 and 3. Chapter 2 highlights a core truth: humans genuinely love music and are deeply passionate about it. Historic and modern-day evidence show that this obsession runs deep, affecting personal lives, societies, religions, politics, and culture. Music's persistent presence across time and places strongly suggests that humans naturally possess musicality, the ability to engage with and respond to music, and that it drives our obsession with it.

Chapter 3 expands on this, exploring how similar, even identical, musical scales are across different cultures and periods. These scales usually start with recognizing octaves on instruments and similar methods of dividing the tonal space within those octaves. These commonalities are not coincidental; they reveal underlying biological, neuroscientific, and physical principles that influence how we perceive and enjoy music.

Together, these facts offer a compelling motive to dive deep and explore. As we examine the central question behind our musical passion, we find a wide range of literature across disciplines like biology, physics, chemistry, and neuroscience. However, much of it is either too technical for the general public or too shallow to be satisfying. Additionally, the information is scattered and often lacks coherence and organization. Like many readers, I've struggled to understand the main reasons for our deep love of music until I realized what was missing and incorporated that into my knowledge base.

What's needed is a holistic, start-to-finish approach, precisely what this book aims to provide. Engaging music is a highly complex process that involves hearing, perception, cognition, emotional response, and physical movement. To understand this complexity, we break the subject down into what I call a phased-naturalism framework: five nearly independent, sequential stages representing different brain activities, from sound production to the feeling of pleasure. These phases are:

1. Music generation
2. Hearing
3. Preprocessing
4. Perception